Textile Chemistry

Theory, Technology, Equipment

Textile Chemistry

Theory, Technology, Equipment

A.P. Moryganov

NOVA SCIENCE PUBLISHERS, INC.
COMMACK, NY

Creative Design: Gavin Aghamore
Editorial Production: Susan Boriotti
Art Director: Maria Ester Hawrys
Assistant Director: Elenor Kallberg
Graphics: Frank Grucci
Manuscript Coordinator: Phyllis Gaynor
Book Production: Joanne Bennette, Michele Keller
Christine Mathosian and Tammy Sauter
Circulation: Cathy DeGregory and Annette Hellinger

Library of Congress Cataloging-in-Publication Data
available upon request

ISBN 1-56072-439-0

6080 Jericho Turnpike, Suite 207
Commack, New York 11725
Tele. 516-499-3103 Fax 516-499-3146
E-Mail: Novascience@earthlink.net

Printed in the United States of America

CONTENTS

PART II CHEMISTRY, PHYSICS AND PHYSICAL CHEMISTRY OF DYES AND TEXTILE AUXILIARIES

PART III THE THEORETICAL BASES OF CHEMICAL TECHNOLOGY OF TEXTILE MATERIALS

PREFACE

It is inconceivable that modern technologies for producing textiles can do without using various natural, artificial and synthetic materials. The properties of all these materials influence greatly the methods of manufacturing textiles and the final quality of textile goods. That is why the questions dealing with synthetic fibers, leathers, fur, dyes, and textile auxiliaries are also discussed.

It seems to us that our approach to the representation of the problems of textile chemistry and of close-related fields of science, technique and technology must be interesting to various specialists.

A.P. Moryganov

Part I

Chemistry and Physics of Natural and Synthetic Fibers

A NEW HEAT-RESISTANT, HARD TO IGNITE FIBER "TOGILEN" WITH HIGH HYGROSCOPICITY AND DYEABILITY

V.M. Pozdnyakov, K.E. Perepelkin, S.A. Baranova,
Zh.Yu. Koytova, V.I. Vakhrushev

The work is devoted to the fiber and thread "Togilen" that has been created at the Leningrad Scientific Research Institute Khimvolokno. Togilen is shown to posses a unique combination of properties: high fire-resistance, sorption characteristics that are close to those of viscose threads and wool, and considerable durability. The good dyeability characteristics of the fiber open up the possibility to produce brightly colored materials meeting high aesthetic requirements.

The combination of new fiber's properties permits the use of Togilen for the production of special purpose fabrics and for the creation of the materials for protective clothes with high hygroscopisity.

Heat-resistant and hard to ignite fibers and threads (terlon, phenilon, arymyd, oxalon, etc.) and the textile materials made from them are widely used for many purposes. These fibers are used in the manufacture of goods to protect people from high temperatures, open flames, and molten metal sparks [1, 2]. But the application of these fibers in special industrial clothing, underclothing, and footwear is often limited by their lack of hygroscopisity. As the sorption curves (Fig. 1) demonstrate [3, 4], the hygroscopisity of terlon and arymyd is much less than that of cotton, wool, and viscose fibers. This low hygroscopisity reduces the hygienic and the wearer comfort characteristics of such heat-resistant and hard to ignite fibers and thereby restricts the possibilities of their use. Heat-resistant and hard to ignite fibers are often processed in blends with wool and other hygroscopic fibers. This results in the lowered resistance of such manufactured products to high temperatures.

At the Leningrad Scientific Research Institute Khimvolokno, the fiber and thread called "Togilen" were created. Togilen belongs to the group of heat-resistant and hard to ignite polymers [5]. It is made of industrially produced monomers: dichloroanhydrides of phthalic acids and one diamine from the benzimidazole series. Togilen is characterized by resistance to heat of up to 300-350^{0}C and does not burn while in contact with an open air flame. The properties of Togilen fibers and threads are described in Table 1 [1, 2].

Table 1.

Properties	Complex thread	Fiber
Specific breaking load, MPa	300-500	250-400
Oxygen index, %	45	45
Wetness (%) at a relative air humidity of 65%	13-14	13-14
Elongation at break, %	15-20	20-30
Resilence module (acoustic), MPa	18000	18000
Tensile strength, %		
at snarl	80	-
at loop	90-100	-
Shrinkage after boiling and consecutive drying, %	2	2
Shrinkage in air (%) at temperature, ^{0}C		
300	0	0
400	1	1
500	5-7	5-7
Strength preservation after thermoaging at 300^{0}C for 100 hours, %	80-90	80-90

The main distinction of Togilen from other fibers in the same group is its high humidity absorption (fiber humidity equals 13-14% in standard conditions) [3, 4]. The data, describing water steam sorption and desorption by Togilen in comparison with other hygroscopic threads (Fig. 1), reveal that in hygroscopic properties, Togilen exceeds wool and cotton and is close to viscose threads. Like these fibers, Togilen displays sufficient hysteresis in sorption characteristics.

Three typical parts are identified on the Togilen sorption-desorption isotherms, and these parts are also characteristic of other hydrophilic fibers. In moisture adsorption and

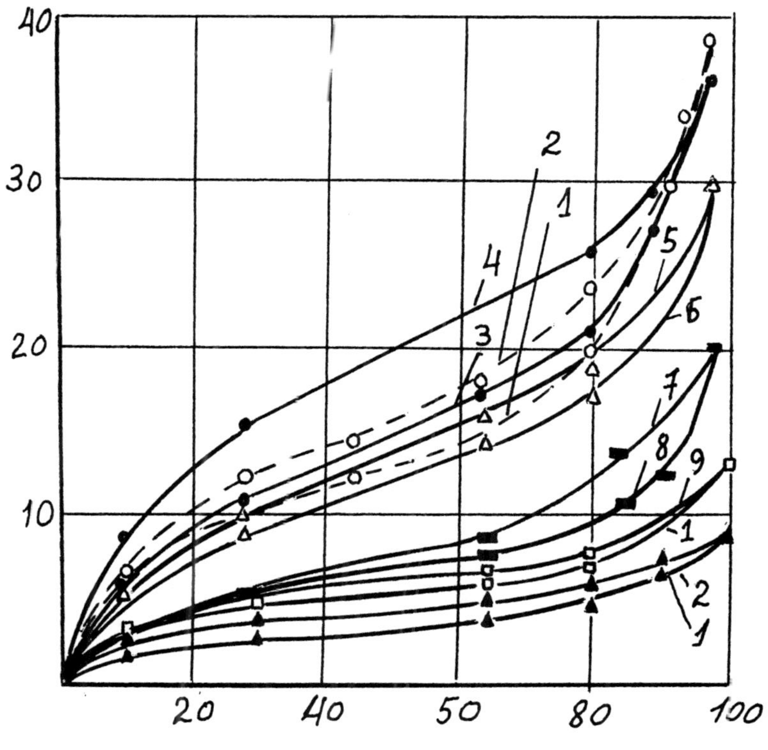

Fig. 1. Thread humidity in sorption (1, 3, 5, 7, 9, 11) and desorption (2, 4, 6, 8, 10, 12) processes as a function of air humidity : 1, 2 - viscose thread; 3, 4 - Togilen; 5, 6 - wool; 7, 8 - cotton; 9, 10 - arymyd; 11, 12 - terlon.

X = φ, % (air humidity); Y = W, % (thread humidity).

desorption rate, Togilen is similar to viscose threads. In equilibrium wetness at different values of relative air humidity, it is close to wool and viscose fibers (Table 2).

Research into the swelling of Togilen in water [3, 4] has shown that in this parameter, Togilen is similar to wool (Table 2).

Table 2.

Properties	Textile thread			
	Togilen	viscose	wool	cotton
Breaking load, cN/tex	30-55	16-20	11-14	25-35
Elongation at break, %	15-20	17-25	25-35	7-8
Equilibrium humidity (%) at relative air humidity, %				
65	16-22	11-14	13-15	7-8
95	29-31	27-33	25-40	20-25
Swelling in water, %	44	66-115	42-49	40-75
Relative wet tenacity, %	30-40	40-50	70-80	110-115

As can be seen from Table 3 [3, 4], the mechanical properties of Togilen fibers and threads depend on their water content. The values of breaking load decrease gradually as air humidity (which defines the quantity of water adsorbed by Togilen) increases. It is indicative of the plastifying influence of water on the fiber structure. The maximal loss of strength is about 50% from the tensile strength of the fiber pattern tested in standard conditions. Elongation at break rises solely at high values of relative air humidity and does not change at lower values.

Table 3.

Relative air humidity at 20^0C	Thread humidity, %	Specific breaking load, cN/tex	Breaking load variation coefficient, %	Elongation at break, %
0	0	43.7	1.7	15.8
10	6	38.7	3.5	15.7
32	10.1	32.3	1.4	15.5
58	-	30.4	1.5	16.4
66	15.9	29.5	1.6	16.5
81	19.3	27.2	1	17.9
92	27.5	23.2	1.1	19
98	-	20.3	1.3	19.7
100	35.6	16.6	7	20.8

In the course of the research into the kinetics of the alteration of Togilen thread tensile characteristics in water, it was established [3, 4] that this thread looses its tenacity appreciably, after having become wet, like viscose (Table 2). Intensive diminution of the breaking load takes place only during the first 5 hours. The cause of this process obviously lies in water molecule diffusion and quick structural changes in the fiber. The structural alterations, taking place in the fiber under the influence of water, then become slower, and result in less intensive and more durable breaking load variations after the 5 hour period. After 25 hours have passed, the tensile strength equilibrium is achieved. When the process of elongation at break alteration is more prolonged, equilibrium is observed only after 500 hours of thread exposure to treatment in an aqueous medium. Up to that time, elongation doubles. Drying of the thread leads to the almost complete restoration of its properties.

The majority of heat-resistant aromatic fibers are yellowish, and the color derives from the presence of chromophoric groups in the polymer structure, as well as from the presence of the products of oxidation and the admixtures.

The main functional groups of heat-resistant aromatic fibers are: amide, imide, oxazole, benzimidazole, etc. That is why these fibers can be dyed by acid dyes. But the high packing density of the above-mentioned polymer structure considerably hampers the dyeing process in which acid and basic dyes are used. The yellow color of the fiber also hinders the production of bright, rich colorings.

The Togilen fiber, which has the most friable permolecular structure (in comparison with other synthetic fibers) and strongly basic tret-nitrogen in imidazole group, is characterized by a heightened affinity to acid dyes and can be dyed brightly in red, green, and blue.

Among hard to ignite, heat-resistant fibers, only PBA fiber, ("Celanise", USA) with a humidity of 10-15% in standard conditions, can be compared with Togilen. The former possesses similar mechanical and thermal properties to Togilen [2, 5, 6, 7]. But PBA fiber is produced from the carcinogenic material (diaminobenzidine) whose synthesis is multi-staged, and the fiber itself is difficult to stain.

Thus, Togilen is notable for its unique combination of properties: high fire-resistance (similar to those of the other heat-resistant threads), good adsorption characteristics (close to those of viscose threads and wool), and tensile properties exceeding by two or three times the same indices of well-known hydrophilic fibers. Having similar swelling properties in water to those of wool and cotton, Togilen (like viscose threads) upon becoming wet loses its tenacity but completely restores it upon becoming dried. The high dyeability of the fiber (resulting from its friable structure) and its high adsorptivity make it possible to develop materials that are brightly colored and are capable of satisfying high aesthetic requirements.

This combination of properties opens wide possibilities for using Togilen threads for the production of fabrics for special purposes and for the creation of materials with the high hygroscopisity required for protective cloths.

References

1. Nachinkin,, P.E. Mikhailov and K.E Perepelkin,, *Khim. Volokna,* (6) (1990) 3 (in Russian).

2. Perepelkin, *Khim. Volokna,* (4) (199) 27 (in Russian).

3. Zh.Yu Koitova et al., *Thesis of the report on the All-Union Scientific Conference "Problems in the Modification of Natural and Synthetic Spinnable Materials".* (Moscow, 1991) 143 (in Russian).

4. Zh.Yu Koitova., K.E Perepelkin and V.M.Pozdnyakov, *Khim. Volokna,* (6) (1992) 43 (in Russian).

5. M.M. Gelmont et al., in *Collected Articles: Chemical Fibers, Fibrous Materials for Technical Purposes* (Moscow, 1990) 114 (in Russian).

6. R.R Jackson,. *Textiles Res. J.,* **48** (6) (1978) 314.

7. R.N De Martino,. *Chemiefaseru Textilindustrie*, **35/87** (1) (1985) 35.

Scientific Research Institute "Khimvolokno",

St. Petersburg

A NEW NON-POLLUTING FIRE RETARDANT SYSTEM FOR POLYMERS

S.M.Lomakin and G.E.Zaikov

N.Semenov Institute of Chemical Physics, Russian Academy of Sciences, Moscow, 117997, Russia

A new-type non-polluting system for reducing combustibility of polymeric materials based on polymeric organic carbonizers (compositions including polyvinyl alcohol) is suggested for Nylon 6.6. This system promotes reactions of high-temperature cross-linking (synergistic carbonization) and formation of carbonized structures. Combustion of polymeric compositions is studied on a cone-calorimeter. The results support the hypothesis underlying the suggested approach.

One of the key problems in utilizing and producing contemporary polymeric materials is their ecological safety. Combustion of polymers and incorporation in them of new compounds (fire retardants) enhancing their fire resistance is directly related to pollution of the ambient atmosphere. At present, organic compounds containing a high percent of haloids (Cl and Br), phosphorus, and heavy and transition metals (Zn, V, Pb, and Sb) are used as fire retardants. All these additives pollute the ambient atmosphere not only when the materials are burning but when they are produced as well. Changing from the commonly known polluting fire retardants to fundamentally new ecologically safe chemical systems reducing polymer combustibility is considered by polymer manufactures as one of the major problems.

Polymer combustibility can be reduced as follows:

1) by modifying the mechanism of polymer pyrolysis to obtain non-volatile or incombustible products that dilute the oxygen-containing atmosphere in the combustion zone;

2) by inhibiting the active radicals entering into chain-radical reactions in the gas phase;

3) by producing a coke layer on the surface of a burning polymeric material that influences the heat and mass exchange processes and thereby suppresses the combustion process.

We concentrated on item 3) in our study. Coke formation in burning of a polymer and its fire resistance are directly correlated. This stems from the fact that coke (carbonaceous layer) retards the flame spreading over the polymer surface and functions as a thermal barrier coating the polymer material. The ability of polymers to form a coke layer can be enhanced by some chemical additives as well as by changing their molecular structure. At present, there are several systems that reduce combustibility of polymeric materials due to formation of carbonaceous layer. Of them most commonly used are the systems containing organo-phosphorous compounds. However these fire retardants only slightly rank below the halogen-containing organic fire proofing compounds in their polluting effect. In the present work we suggest for the first time a polymeric additive based on polyvinyl alcohol (PVA) as a fire retardant (coke forming agent) for polymers. This additive forms conjugated systems (subsequently, aromatic structures) when heated that actively participate in coke formation from the polymeric material at the last stage of high-temperature pyrolysis.

We proposed a non-polluting system based on polyvinyl alcohol (high-temperature polymeric carbonizer) for reducing combustibility of Nylon 6,6.

Table 1.

Products of Thermal Degradation of PVA at 240 °C [1]

Products	wt %	Products	wt %
Water	33.400	3-pentanole-2	0.190
CO	0.120	3,5-heptadienone-2	0.099
CO_2	0.180	2,4-hexadienol	0.550
C_1—C_1 hydrocarbons	0.010	Benzaldehyde	0.022
Acetaldehyde	1.170	Acetophenone	0.021
Acetone	0.380	2,4,6-octatrienol	0.110
Ethanol	0.290	3,5,7-nonatrienol	0.020
Benzene	0.060	Unidentified	0.082
Masityl oxide	0.760		

Thermal decomposition of polyvinylalcohol proceeds in two stages. At the first stage, that starts at 200 °C, dehydration is the dominating process. The remaining polymeric product consists mainly of conjugated unsaturated structures of the polyene type. At the second stage (450 °C) the polyene structures decompose further to produce a great amount of hydrocarbons and a small amount of a carbonaceous residue. A detailed mechanism of thermal decomposition of PVA was suggested by Tsuchiya and Sumi [1]. At 25 °C, a water molecule is eliminated from the polymer link to produce an unsaturated polyene structure:

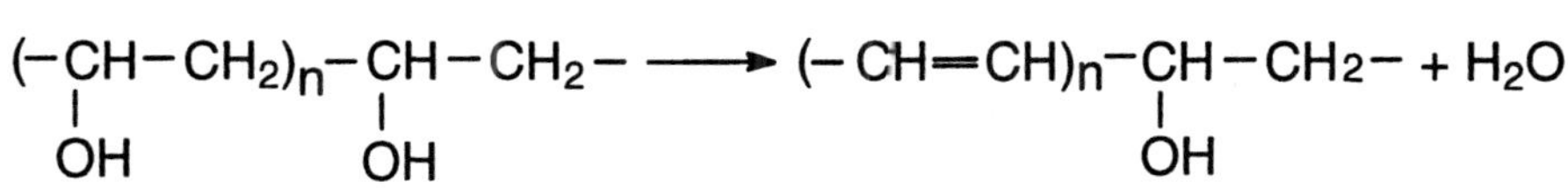

Fission of the C-C bonds leads to formation of terminal carbonyl groups. For example, in the course of decomposition the number of therminal aldehyde groups increases:

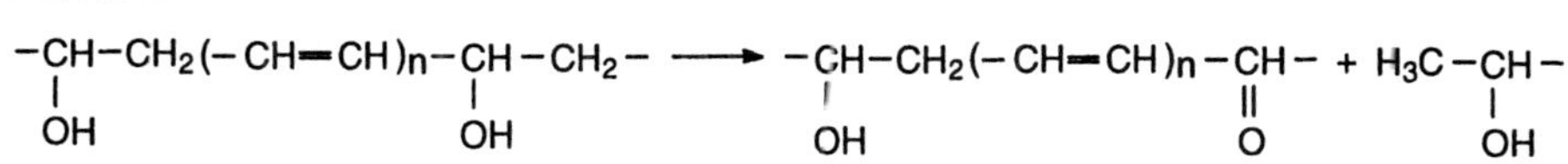

At the second stage of PVA pyrolysis the volatile pyrolysis products comprise predominantly hydrocarbons: n-alkenes and aromatic hydrocarbons (Table 1) [1].

Thermal decomposition of PVA in the presence of oxygen can adequately be described by two-stage decomposition mechanism with the only exception. Oxidation of the unsaturated polymeric product formed by the dehydration reaction produces ketone groups in the polymeric links. These groups promote dehydration of neighboring vinyl alcoholic groups and form an unsaturated conjugated ketone polymeric structure [2]. The first stage of thermal oxidative PVA decomposition resembles decomposition in vacuum. At temperatures from 260 to 280 °C, the kinetics of thermal oxidative PVA decomposition satisfies the equation for a second-order reaction. The activation energy for this reaction is 53.6 kcal/mol [2].

Changes in the IR spectra of thermally treated PVA samples were scrutinized in [2]. At 180 °C in air, a new absorption band arises at 1630 cm^{-1} (vibrations of the isolated C-C bond), 1650 cm^{-1} (C-C vibrations in conjugated dienes and trienes), and 1590 cm^{-1} (C-C vibrations in polyenes). The intensity of stretching vibrations of carbonyl groups in the range 1750-1720 cm^{-1} increases, although the rate of this increase is lower than that for polyene bonds at lower temperatures.

At temperatures above 180 °C, the oxidation rate rises after an initial induction period. The authors of [2] believe that the presence of a small benzene amount in the products of PVA decomposition is direct proof of cross-linking reaction in polymer proceeding via the Diels-Alder mechanism:

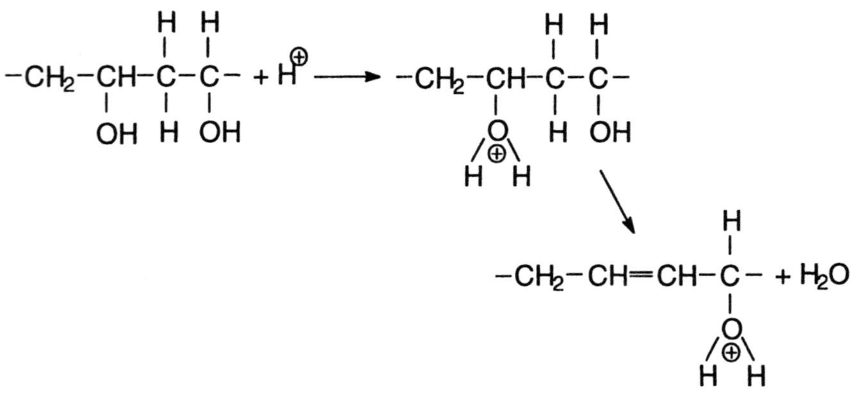

Purely aromatic structures are formed in the solid residue, which follows from the IR spectral data [2].

The dehydration reaction catalyzed by acids accelerates formation of the system of consecutive conjugated double bonds (a), and Diels-Alder addition of conjugated and isolated double bonds in different polymer chains may lead to intermolecular cross-linking, carbonization, and graphitization (b):

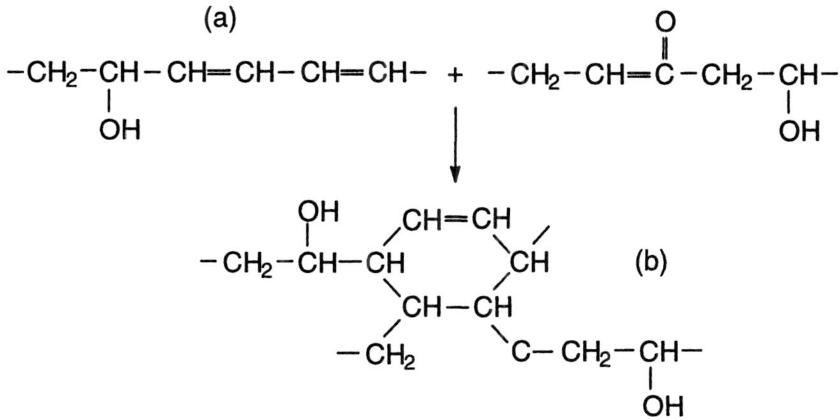

In contrast of PVA, thermal decomposition of Nylon 6,6 in an inert medium at temperature above 300 °C progresses to the end with no residue [3,4]. A great variety of the decomposition products including various hydrocarbons, cyclopentanone, water, CO, CO_2 and NH_3 points to the complex nature of the decomposition process. The comprehensive studies resulted in the commonly adopted mechanism of thermal decomposition of aliphatic polyamides [5]

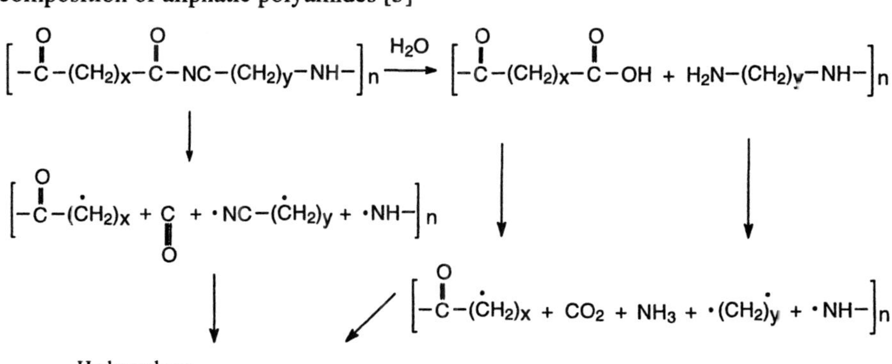

Hydrocarbons,
Cyclic ketones, ethers, nitriles, coke

1. Hydrolysis of amide bonds starts normally at a temperature which is lower than the temperature corresponding to the beginning of appreciable decomposition.

2. The next stage of the process is a homolytic rupture of the C-C, C-N, and C-H bonds (the temperature at which significant decomposition starts).

3. Then cyclization and homolytic decomposition of the reaction products take place.

4. The final stage in the suggested mechanism is secondary reactions producing CO, NH_3 , nitriles, and a carbonaceous residue

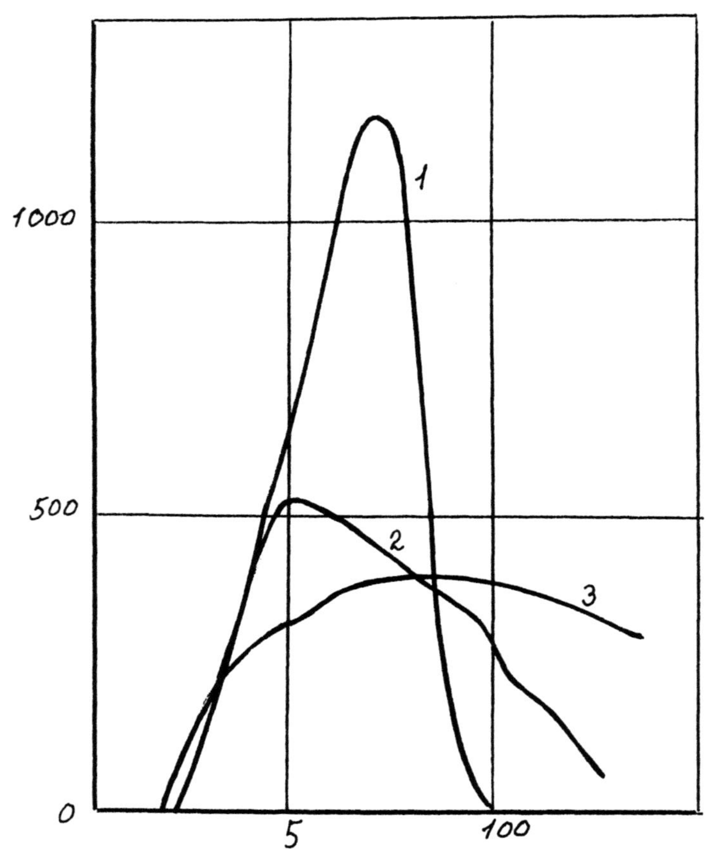

Figure 1. The mechanism of synergic carbonization of the Nylon 6.6.-PVA composition

The idea of creating Nylon 6,6-PVA compositions is based on feasibility of high temperature (acid-catalytic) PVA dehydration. This reaction is catalyzed by the acidic molecules liberated at the first stage of PVA decomposition. Further on, the products of these reactions undergo intermolecular cross-linking and eventually form the carbonaceous residue. We call this mechanism "synergistic carbonization" (Fig.1). The term synergism in this case means not only the combined effect of PVA and Nylon on carbonization of the system but also the fact that the final coke yield exceeds substantially the appropriate parameter for individual PVA and Nylon 6,6.

As well recognized, polyamides are poorly compatible with other polymers because of their strong hydrogen bonds. Previously, the authors of [6] have studied compatibility of Nylon 6,6 polyvinylacetate, and PVA. Compatibility was assessed from the decrease in the melting point. Investigations have shown that Nylon 6 - PVA mixtures are partially compatible. The term compatibility in this case means that the mixture prepared is a fine structure of the domain type which can be obtained by simple pressing, rather than the true thermodynamic compatibility.

The next step in our study of the Nylon-PVA system was an attempt to replace PVA by PVA oxidized in aqueous $KMnO_4$ solution. This approach was based on the unique fire-suppressing properties of PVA oxidized by $KMnO_4$ (PVA-$KMnO_4$). This was ascertained experimentally (using the cone-calorimeter apparatus) that the heat release rate reduced substantially and the induction period of self-ignition of the PVA-$KMnO_4$ system increased as compared to PVA (Table 2).

Table 2.

Cone-Calorimetric Data for Polymeric Compositions

Material, initial thermal flux, kW/m^2	Sample weight, g	Coke yield, wt %	Self-ignition period, s	Maximum heat release, kW/m^2	Total heat release, MJ/m^2
PVA, 20	47.6	8.8	39	255.5	159.6
PVA, 35	28.3	3.9	52	540.3	11.3
PVA, 50	29.2	2.4	41	777.9	115.7
PVA-$KMnO_4$, 20	27.9	30.8	1127	127.6	36.9
PVA-$KMnO_4$, 35	30.5	12.7	774	194.0	103.4
PVA-$KMnO_4$, 59	29.6	9.1	18	305.3	119.8
PVA(100°C)-$KMnO_4$, 20	31.1	16.3	303	211.9	124.5
PVA(200°C)-$KMnO_4$, 20	35.9	25.7	357	189.0	91.1
Nylon 6.6, 50	29.1	1.4	97	1124.6	216.5
Nylon 6.6-PVA (8:2), 50	26.4	8.7	94	476.7	138.4
Nylon 6.6 - PVA (8:2) $KMnO_4$, 50	39.1	8.9	89	399.5	197.5

Until recently, polymeric redox systems ($KMnO_4$) received little attention of investigators. The authors of [7, 8] studied oxidation by potassium permanganate of a polymer containing secondary hydroxyl groups (PVA). They supposed that PVA oxidation in an alkali solution of $KMnO_4$ can proceed via two intermediate complexes, (1) and/or (2) [8]:

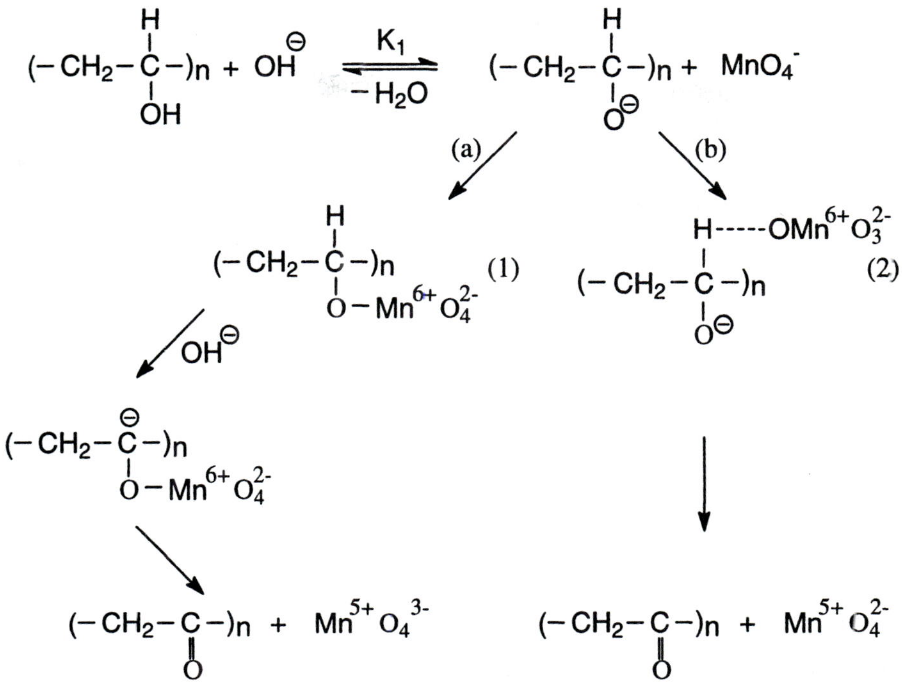

Reactions (a) and (b) lead to formation of the end product -polyvinylketone (3). Polyvinylketone was isolated and identified using microanalysis and spectroscopy (IR and UV spectral data) [9].

Experimental

Materials. We used 99% hydrolyzed PVA (M_w =86000); Nylon 6,6 (Sci.Plo.Prod.); and class R potassium permanganate (BA Chemical Ltd.).

Experimental procedure. Samples of polymeric mixtures (Nylon 6,6-PVA/PVA-$KMnO_4$) were prepared in a laboratory mixer at room temperature (10 min) and pressed at a pressure of 10 t/m^2 and temperature of 220-240 °C.

Polyvinyl alcohol was oxidized by $KMnO_4$ in aqueous solution. Aqueous solution of PVA (10%) was prepared at 90 °C in a reactor 2 l in volume. Potassium permanganate (5 wt % of the initial PVA weight) was added to the hot PVA solution. After completion of the fast reaction (1.5-2.0 min) the solution became dark brown. Then it was cooled to room temperature. The resulting viscous solution was evaporated under vacuum in a rotor vaporizer at a temperature of 50 °C to produce an elastic residue. Then this residue was dried at 120 °C for 24 h until solid dark brown material was obtained. This material was crushed in a ball mill to the state of a fine powder.

The combustion parameters of polymeric materials were measured on a conecalorimeter (NIST-USA). Samples pressed in the form of discs (35 mm in diameter) were subjected to external constant thermal fluxes of 20, 30, 35 and 50 kW/m^2.

Results and discussion

Preliminary experiments with PVA and PVA-$KMnO_4$ materials conducted on the cone-calorimeter at different initial thermal fluxes clearly indicate that fire proofing properties of PVA-$KMnO_4$ are much better than those of PVA (Table 2). The data listed in Table 2 show a considerable increase in the amount of carbonaceous residue and nearly two-fold reduction of the maximum heat release rate (R.H.R., kW/m^2) in burning PVA-$KMnO_4$ samples as compared to pure PVA samples. Even at an initial heat Flux of 50 kW/m the coke yield for PVA-$KMnO_4$ was 9.1 wt %. The self-ignition periods for PVA-$KMnO_4$ are naturally longer than those inherent in PVA (Table 2). Such a high fire resistance of PVA-$KMnO_4$ can be ascribed to the fact that PVA-$KMnO_4$ (polyvinylketone-containing PVA structures) can enter into reactions as a neutral (ketone structure) (1) or enol-type (monobasic) (2) structure [9]:

(1) (2)

(Fig.2)

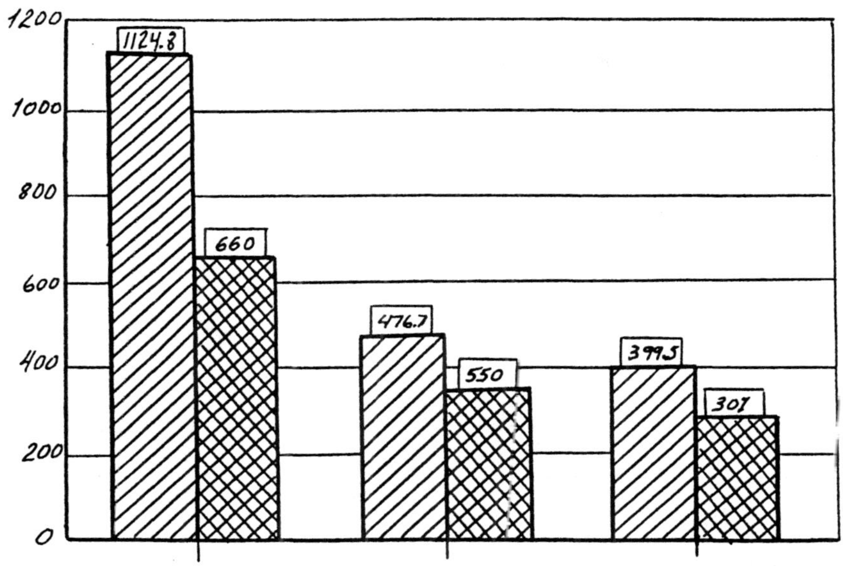

Figure 2. Time histories of the heat release rate in sample burning: (1) Nylon 6,6, (2) Nylon 6,6-PVA (80:20), (3) Nylon 6,6 - PVA-$KMnO_4$ (80:20); the initial heat flux is 50 kW/m^2

The IR and UV spectral data [6] suggest that polyvinyl ketones forming coordination compounds with some metal ions (Cu^{2+} , Ni^{2+} , Co^{2+} , Cd^{2+} , and Hg^{2+}) have predominantly the enol type (2) structure under the reaction conditions. Based on that the following structure of the polymer "intermolecular complex" was proposed:

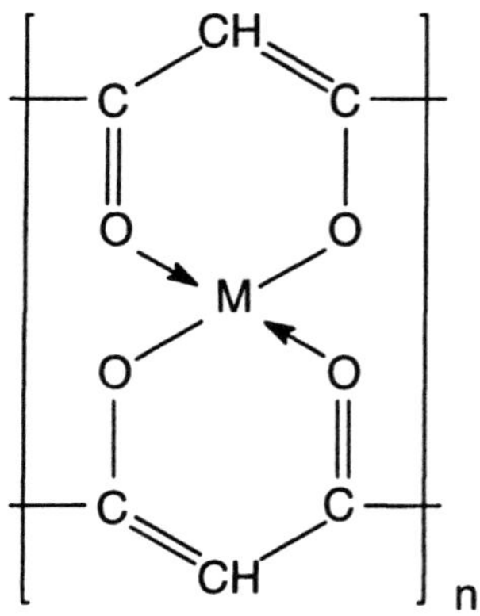

M is metal

The results of the elemental analysis of PVA-$KMnO_4$ point to the presence of 1.5 wt % of Mn in the polymer chelate structure. Thus, we may suppose that this catalytic content of the Mn-chelate structures in the polymer matrix promotes fast high-temperature carbonization and graphitization processes. Cone-calorimetric measurements performed with two PVA samples mixed with $KMnO_4$ in the solid phase (5 wt %) at temperatures of 100 and 200 °C demonstrate unequivocally the superiority of liquid-phase (homogeneous) introduction of $KMnO_4$ in PVA structure (Table 2).

Tests performed at 50 kW/m^2 with Nylon 6,6 and Nylon 6,6-PVA (80:20) mixtures supported the idea of synergistic carbonization (Table 2, Fig.2). (Figure 1 is a typical illustration of the time dependence of the heat release rate in burning polymer samples). Each of the individual polymers (PVA and Nylon 6,6) is less fire-proof than their compositions. The mechanism of synergistic carbonization is presented in Fig 1. The above trend can be traced not only for the maximum heat release but for the rate averaged throughout the sample burning period as well (Fig.3).

PVA-$KMnO_4$ samples exhibit better fire-resistance properties owing to the catalytic effect of the Mn-chelate fragments on the carbonization process, this effect is seen from the maximum and average heat release rates (R.H.R) (Table 2, Fig.3). However, a comparison of the total heat release in burning samples furnishes less satisfactory results (Table 2). Although the experimental results are indicative of reduction of the total heat release in the Nylon 6,6-PVA and Nylon 6,6-PVA-$KMnO_4$ samples exhibit slightly higher total heat release values than do the Nylon 6,6-PVA samples (Table 2). We ascribed this finding to the fact that the Mn-chelate fragments present in the polymer structure catalyze smoldering of polymeric samples. The time of complete burn out of Nylon 6,6-PVA-$KMnO_4$ samples is longer than that of Nylon 6,6 and Nylon 6,6-PVA samples (Table 3). The average heat release rates in burning samples point at the exothermic nature of the smoldering process catalyzed by the Mn-chelate structures (Table 3). Therefore we were able to detect an approximately identical amount of the carbonized residue for the Nylon 6,6-PVA and Nylon 6,6-PVA-$KMnO_4$ samples (Table 2).

Table 3.

Cone-Calorimetric Measurements of the Parameters for Compositions with Nylon 6,6 at an Initial heat Flux of 50 kW/m^2

Composition	Complete burn-out time, s	Average heat release, MJ/kg
Nylon 6.6	312	31.50
Nylon 6.6—PVA (80:20%)	429	25.15
Nylon 6.6— PVA-$KMnO_4$ (80:20%)	747	29.52

Conclusion

Organic polymeric carbonizers (PVA systems) introduced in Nylon 6,6 lower the maximum heat release rate from 1124.6 kW/m^2 (for Nylon 6,6) t0 777.9 kW/m^2 (for PVA) to and increase the coke yield for Nylon 6,6 from 1.4 to 8.7% due to the effect of synergistic carbonization (at the initial thermal flux of 50 kW/m^2).

Experiments conducted on the Cone-calorimeter with Nylon 6,6-PVA-$KMnO_4$ compositions have shown better fire-resistance characteristics by way of example of the maximum heat release rate in burning material that decreases from 476.7 kW/m^2 for the

Nylon 6,6-PVA compositions to 399.5 kW/m^2 for the Nylon 6,6-PVA-$KMnO_4$ composition. On the other hand, introduction of PVA-$KMnO_4$ in Nylon 6,6 gives rise to Mn-chelate fragments that catalyze the solid-phase smoldering processes and increases the heat release value of the Nylon 6,6-PVA-$KMnO_4$ compositions as compared with the Nylon 6,6-PVA composition. This, in turn, affects the total heat released in burning of polymeric materials.

Organic polymeric carbonizers (PVA, cellulose materials and so on) may serve as new ecologically safe systems reducing combustibility of polymers.

References

1. Y. Tsuchiya and K. Sumi, *J.Polym.Sci.,* **7** (1969) 3151.

2. *Polyvinyl Alcohol. Properties and Applications,* Ed. C.A. Finch (London: J.Wiley Sons, 1973).

3. B.G. Achhammer, F.W. Reinhard and G.M.Kline, *J.Appl.Chem.* **1**, (1951) 301.

4. I.Goodman, *J.Polym.Sci.,* **13** (1954) 175.

5. *Thermal Stability of Polymers*, Ed. R.R.Conley, (Marcel Decker Inc., New York, 1970), vol.1

6. Ha Chang-Sik, Lee Won-Ki, Roe Tae-Woo, and Cho Won-Jei, Polymer Bull. 31, 359 (1993)

7. R.M. Hassan, Polymer Intern. 30, 5 (1993)

8. R.M. Hassan, S.A. El-Gaiar, and A.M.El-Summan, Polymer Intern. 32, 39 (1993)

9. R.M. Hassan, M.A. Abd-Alla, and M.A.El-Gahmi, J.Mater.Chem. 2, 613 (1992)

PREPARATION OF FIRE-PROOF POLYETHYLENE TEREPHTHALATE USING NEW FORMS OF THE PHOSPHORUS POWDERY FIRE-RETARDANT

N. S. Zubkova, N. Yu. Borovkov*, M. A. Tyuganova
A. P. Moryganov*

Fire-retarding effect of a powdery fire-retardant (FR) on the base of methylphosphonic diamide and its microcapsules (MC) has been investigated. The MC contain TiO_2 and SiO_2 as binders, content and nature of the binder determining fire-retarding activity of the MC. Fire protection of thermoplasts with the MC and ammonium polyphosphate (APP) is looked upon comparatively. Properties of fire-proof polyester fibers prepared by moulding of the melt have been investigated. Deductions are made how to improve this FR by microencapsulation.

Method of the fire-proof synthetic materials production by introduction of FR into the polymer melt before moulding makes it possible to retain technologies already developed. This method is attractive economically and provides the basis for development of ecologically safe technologies. However, FR to be used for moulding should meet a great deal of requirements. FR should be thermostable and not destruct polymer. These FR should be active enough to prevent ignition at their small concentration in the mixture (10-20% of polymer weight) [1].

An urgent problem of fire-proof carbonizable thermoplasts preparation, e. g. polycaproamide (PCA) and polyethylene terephthalate (PET), is a reduction of volatile products yield. Another important problem is to raise polymer carbonizability because it is an intumescent char layer that makes it difficult for oxygen to diffuse into inner layers of burning polymer [2]. Drastic intensification of PCA and PET charring that entails extinction of the flame takes place in mixtures of the polymers with phosphorus-containing FR. The most powerful fire-retarding compositions are known to be ones containing phosphorus, nitrogen and halogen simultaneously. In such composition a phosphorus-containing ingredient is active mainly in the condensed phase of pyrolysis and burning. As for nitrogen and halogen, they inhibit oxidative radical processes in the gas phase and dilute volatile products of pyrolysis to concentrations at which ignition and self-ignition are impossible.

With due regard for these considerations we took commercial FR T-2 as an object of our investigation. This FR being a product of methylphosphonic dichloride ammonolysis is a mixture of ammonium hydrochloride and methylphosphonic diamide [3]. Commercial samples of PCA and PET with the average molecular weights (MW) 20,000 and 30,000 were used. Fire-proof mixtures were prepared as follows. Polymer granules mixed beforehand with T-2 were melt and thereupon fibers were molded from the melt with a screw extruder. Limiting oxygen index (LOI) tests were carried out according to the Russ. State Standard 12.1.044-89 (Russian analog of the ASTMD 2863) using a Stanton Redcroft apparatus. Thermal degradation was carried out under nitrogen on a Thermal Analyzer Q-1500 (Hungary). Volatile products of pyrolysis were analyzed by stepwise pyrolytic gas chromatography using a "Chrom-5" apparatus (Czechoslovakia).

It has been stated that T-2 decomposes endothermically absorbing 1625 J/g with a peak rate (V_{max}) within an interval of the thermoplasts decomposition (340-410°C). It is noteworthy that such widely used FR as aluminum hydroxide and APP have endoeffects 1030 and 1206 J/g, correspondingly. The value for tetrabromphthalic anhydride is a far less than the above ones (256 J/g at 320oC). Data on T-2 and APP listed in Table 1 confirm T-2 to be a highly active FR for the thermoplasts. The melting point of PCA as well as V_{max} value decrease and char yield increases significantly in PCA mixtures with T-2 (Table 1). These facts are true in the case of PET too.

Table 1

Phosphorus-containing fire-retardants action on properties and thermooxidation of polycaproamide and polyethylene terephthalate

Poly-mer	Fire-retardant	P, %	M.P. oC	TGA data			LOI %	Ab
				t max, oC	V maxa, mg/min	Char at 500oC, %		
PCA	-	0	231	440	17	5	19	-
	25% T-2	3.0	210	390	12	13	27	1.9
	18% T-2 + 3,6%SiO_2	2.7	198	309	7	32	28	2.7
	25% APP		-	370	10	15	24	1.3
PET	-	0	267	470	24	12	20	-
	10% T-2	1.3	265	440	15	16	28	4.1
	10%T-2 + 2.0% SiO_2	1.2	265	425	11	20	28	4.4
	25% APP	3.0	-	450	20	14	26	1.7

a Volatile products evolution rate; b FR activity parameter (the LOI increment per 1% P introduced)

Because T-2 has some drawbacks (low temperature of decomposition onset, melt viscosity reduction), an attempt has been made to combine it with silica, a well-known thickener of polymers melts [4]. Stabilization of melt viscosity occurs when portions of 3.5-4.0% Aerosil (PCA) and 2.0% Aerosil (PET) are added together with T-2. Activity of T-2 combined with silica is higher and V_{max} values decrease from 12 mg/min for PCA mixed with pure T-2 to 7 mg/min for the mixture PCA-FR-silica (Table 1). Combination of T-2 and Aerosil results in changes of the char optical properties as well raising the reflection coefficient value and reducing the transmission coefficient value. Such changes are known to diminish heat loss from burning polymer surface and so contribute to its thermooxidative stability.

However, a problem of unsatisfactory physicomechanical properties of the fire-proof thermoplasts arises when the polymers are being moulded from the melt mixed with T-2 and Aerosil. A source of the problem is appeared to be the partial decomposition of both

T-2 in the melt and the polymer itself due to thermooxidative destruction of the latter. As a result of the decomposition, the MW value decreases. At the same time one observes changes in a quantity of functional groups which is known to be an important parameter representing polymer thermooxidative stability [5]. Our measurements give the quantity of groups equal to 45 and 102 mg/eq for pure PET and for PET moulded with 10% T-2, respectively.

To eliminate the drawbacks a search for new forms of T-2 was undertaken. MC of T-2 with the aramide wall were prepared and their fire-retarding effect was investigated [6]. Application of T-2 microencapsulated with 6-10% Nomex as a wall material raised the LOI value of PET up to 28-29%. However, the microcapsules with the 15-20% Nomex wall were less active than pure T-2 owing to reaction of a great part of T-2 in the cores with Nomex itself under conditions of polymer burning [6].

By this work we have been trying to eliminate parasitic influence of an organic material of the microcapsules wall. We have prepared new forms of T-2 including only chemically inert substances as binders, namely titanium and silicon dioxides. These substances are known to be the active quenchers of peroxide and hydroperoxide radicals that die out on a surface of dioxides [7]. Titanium and silicon dioxides are refractory. They have low specific heat capacity and heat conductivity and are used for the production of filled polyamides and polyesters [8].

Two T-2 forms have been prepared to investigate their fire-retarding effect. They are a modified form (MF) and microencapsulated one (MC). The modified form MF1 has been prepared by heating T-2 with titanium tetrachloride in petroleum ether. The yellow product has been kept under water vapors overnight and dried at 110°C. The MF1 is a white powder and contains 29% water-insoluble residue corresponding to an equimolar T-2-$TiCl_4$ adduct. The MF2 containing polyethylene polyamine (PEPA) has been prepared in the same manner, 5% PEPA being deposited on T-2 from an organic solvent before modification. The MC has been prepared by drying-in-liquid (solvent evaporation) technique [9] using polyethylsiloxane as a continuous medium and ethanol as a solvent. Hydrolyzed tetraethoxysilane at weight proportion T-2:SiO_2 = 10:(1-2) has been used as a binder. The MF and MC particles prepared have diameter of 50-200 micrometer. An idea assumed as the basis of the new forms preparation is to raise the T-2 thermostability through embedding it in an inorganic matrix as well as to prevent harmful interaction between PET and T-2.

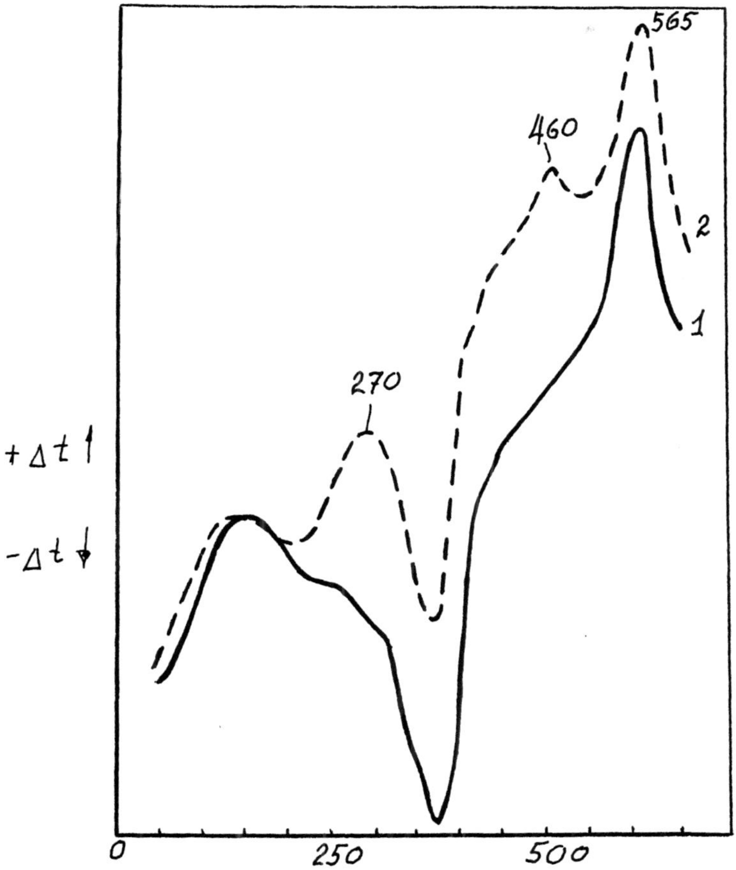

$y = \uparrow + \Delta t$

$\downarrow - \Delta t$

x = temperature, °C

Figure 1. DTG curves for the fire-retardant T-2 modified with titanium tetrachloride: 1. 29% TiO_2 (MF1); 2. 29% TiO_2 + 5% PEPA (MF2)

Figure 1 shows thermogravimetric analysis (TGA) data on the MF. Thermolysis of the MF2 is accompanied by considerable evolving of heat (peaks at 270, 460, 565oC). On the other hand, the MF1 decomposes at 565oC with small exoeffect. As for the MC, TGA does not show noticeable dependence of their thermostability on silica content. When silica content is raised from 9 to 17%, the FR activity coefficient value (A) increases from 3.5 to 4.2 that is nearly equal to the A value for PET mixed with pure T-2. The A values for the PET-MF mixtures are less favorable (Table 2).

Table 2

Thermal analysis data on the T-2 forms and polyethylene terephthalate mixed with them

Binder element	Water-insoluble residue %	P, %	T-2 forms		RET + 10% T-2 form			
			t max, °C	V max, mg/min	t max, °C	Vmax, mg/min	LOI, %	A
Si	9	11.0	335	7.1	435	14.0	26.3	3.5
	12	11.0	332	6.9	430	14.0	26.6	3.8
	17	9.4	330	6.7	450	12.7	26.7	4.2
Ti	29	9.8	350	7.0	450	11.0	26.1	3.8
	34[a]	8.4	327	7.3	435	11.4	25.7	3.8

a With 5% PEPA

Figure 2 shows TGA data on decomposition of the PET-MC mixtures, the MC having various contents of silica. The mixtures with the MC having 9 and 12% silica (curves 1, 2) have V_{max} value 14 mg/min at polymer conversion extent 45-50%. When silica

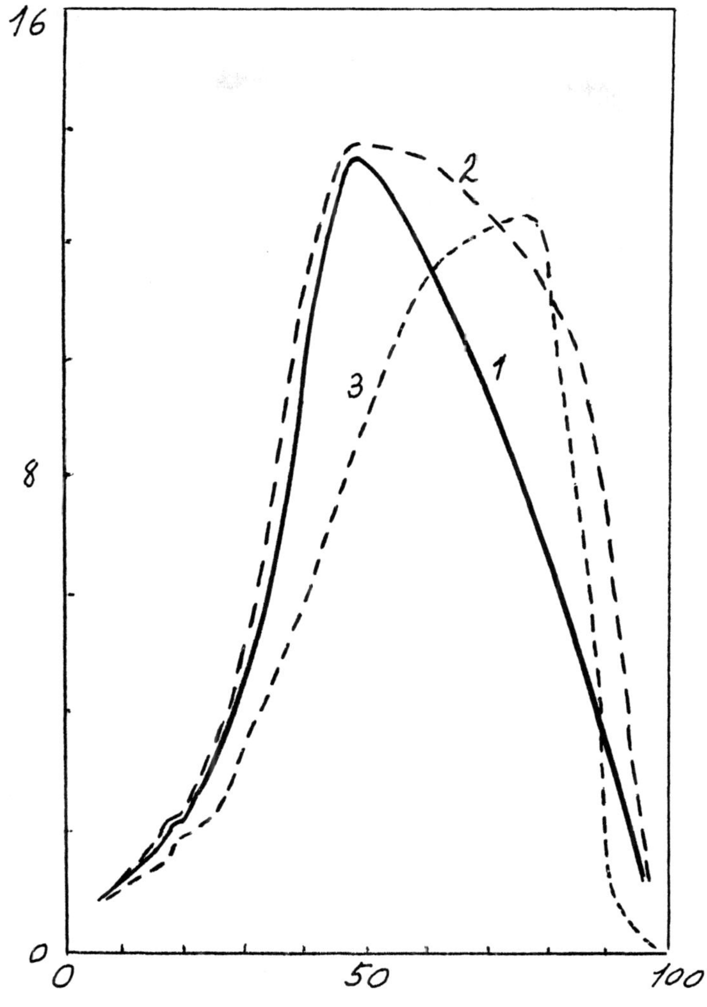

Figure 2 Dependence of volatile products evolution rate on the conversion extent for polyethylene terephthalate mixed with T-2 microencapsulated in silica (MC) at silica content 9 (1), 12 (2) and 17% (3).

content raised up to 17% (curve 3), thermostability of the PET mixture increases significantly, V_{max} being at the 80% conversion extent.

Y = 2W/2T(mg/mik)

X = (1-W) 100%

Application of T-2 in the silica MC does not solve radically the problem of melt viscosity change. The problem remains because of porosity of the silica matrix T-2 embedded in. But now one can suppose physicochemical properties of polymer that should be used as a wall material for MC being stable in the thermoplast melt. The properties are as follows: 1. thermostability level not less than 380-400oC; 2. stability to acid hydrolysis at high temperatures; 3. ability to form films of fine porosity; 4. high adhesion to polar T-2 surface; 5. high solubility in polar media. In the light of the aforesaid, our next step will be investigation of fire-retarding effect of T-2 covered by polymers with purely inorganic main chains of macromolecules.

References

1. *Polymeric Materials with Low Flammability.* Khimiya, Moscow, 1986 (in Russian).

2. M. A. Tjuganova *Theoretical Problems of Fiber-forming Polymers Fire-proofing and Fire-proof Textile Production Technology.* Thesis. Moscow, 1988 (in Russian).

3. R. Raetz *Methylphosphonic Diamide.* J. Am. Chem. Soc., *77*, 4170 (1955).

4. M. T. Bryk *Destruction of Filled Polymers.* Khimiya, Moscow, 1989 (in Russian).

5. A. Angelova, V. Mincheva, S. Voinova and D. Dimitrov *Influence of Stabilizers on Thermooxidation of Polyethylene Terephthalate. Khim. volokna* (Russ. Chem. Fibers), N 3, 19 (1978) (in Russian).

6. N. S. Zubkova, M. A. Tjuganova, N. Yu. Borovkov and A. P. Moryganov *Preparation of Fire-proof Textile Fibers using Fire-retardant T-2 in the Microencapsulated Form. Khim. volokna* (Russ. Chem. Fibers), N 3, 19 (1995) (in Russian).

7. G. P. Gladyshev, Yu. A. Ershov and O. A. Shustova *Stabilization of Thermostable Polymers.* Khimiya, Moscow, 1979 (in Russian).

8. A. A. Konkin *Carbonized Fibrous and Other Heat-resistant Materials.* Khimiya, Moscow, 1974 (in Russian).

9. R. Arshady *Microspheres and Microcapsules, a Survey of Manufacturing Techniques.* Part III. Solvent Evaporation. Polymer Eng. and Sci., *30*, 915 (1990).

State Textile Academy, Moscow;

**Institute of Non-Aqueous Solutions Chemistry,*

Russian Academy of Sciences, Ivanovo, Russia

HARD TO IGNITE MATERIALS CREATED FROM TANNING INDUSTRY WASTES

M.A.Tyuganova, N.S.Zubkova, T.A.Belousova,
N.I.Konstantinova, Yu.K.Noganovskiy

The present work is devoted to the production of hard to ignite materials from tanning industry wastes. It has been shown that the introduction of melamine into the collagen dispersion obtained from tanned and non-tanned leather wastes results in an increase in the fire-resistant properties of these materials. The specificity of the melamine fire-resistant function has been investigated for materials containing collagen. The fire-resistant materials thus obtained can be used for interior decoration.

The formation of a large quantity of wastes (up to 50-60%) as a result of animal hide processing into high-quality leathers is a peculiarity of the tanning industry. Not long ago, the effective cryo-structurizing method for tanned and non-tanned waste processing has been created which enables the preservation of a valuable natural protein - collagen. According to this method, tanned leather wastes are dispersed and non-tanned wastes are subjected to two different treatments: first, with an alkaline salt solution and secondly, with acids (acetic or phosphoric). After the acidic treatment, a homogenization operation takes place up to the production of a fine dispersion. This collagen dispersion is thereupon put into metal pans and kept at -40 - -60^{0}C for 10-12 hours. Then the material is unfrozen, dried, and ironed in order to obtain a smooth surface. The described method was created in a tannery in Kursk, Russia. The materials thus produced possess high hygienic properties and a beautiful appearance. Nevertheless, this method is not widely known because of the high combustibility of the materials.

The difficulty of solving the problem of lowering the combustibility of collagen materials is connected with the specificity of leather protein thermal decomposition and with the complexity of chemical and physicochemical processes taking place during the thermolysis and burning of this material.

Among the retardants of burning of carbonizing polymers (collagen belongs to such polymers), phosphorus-containing compounds are the most frequently used [1, 2]. But low efficiency in retarding gaseous-phase reactions during burning is a shortcoming of the retardants of this group. At present, the possibility to use nitrogen-containing compounds, especially melamine and its derivatives, is attracting more and more attention [3, 4]. The advantages of melamine consist in its thermal stability during heating up to 250^0C and its non-toxicity. However, as a rule, nitrogen-containing compounds do not influence thermolysis and burning processes in the condensed phase. Taking this fact into account, the possibility was studied to use a derivative of methylphosphonic acid and melamine as the retardant against burning. The chemical structure of this compound (called A-128), synthesized by the Volgograd Industrial Association Khimprom, can be represented by the following formula:

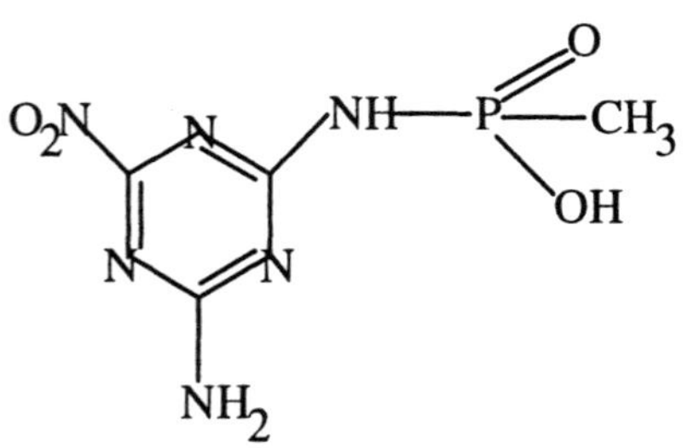

(A-128)

This synthesized retardant and melamine were added to collagen dispersion in the amount of 1-7% of the dry residue mass.

The research into the flame-resistant properties of the chosen retardants was performed in conformity with the adopted Russian standards (GOST 12.1.044-89 and GOST 217930-76).

The thermal analysis of retardant patterns was done using the thermoanalytical complex “Du Pont-9900”. The data obtained are given in Table 1 and in Fig.1 and 2.

Table 1.

N	Additions into collagen dispersion, %	Oxygen index, %	Temperature (^0C) of mass loss equal to 10%	to 20%	to 30%	Specific relative experimental effect of oxidizing, ^0C $mg^{-1}min$	Maximal temperature increment during burning ($T_{max.}$), ^0C
1	H_3PO_4 (2)	33.4	120	329	414	11.7	180
2	H_3PO_4 (2) + melamine (3)	33.1	122	328	407	11.4	210
3	CH_3COOH (2)	44.8	272	322	402	6.9	80
4	CH_3COOH (2)+ melamine (7)	49.6	236	322	392	8.3	75
5	H_3PO_4 (2) + A-128 (7)	37.3	157	321	392	8.4	150
6	Original leather	33.0	142	328	408	12.9	185

As is evident from Table 1, the content of melamine plays the main role in giving flame-resistant properties to leather materials. For patterns 3 and 4, the oxygen index rises from 44.8% up to 49.5% when the melamine content is increased from 3% to 7%. These patterns are also characterized by the lowest heat effects of oxidizing processes.

The use of the phosphorus-containing melamine derivative A-128 leads to an insignificant lowering of the combustibility of the composition. The inadequate efficiency of A-128 is conditioned upon the existence of the chemical bond between methylphosphonic acid and melamine resulting in the decrease of the flame-resistance effect peculiar to nitrogen-containing retardants.

It must be noted that the additions of melamine cause a substantial increase in the temperature at which the composition begins to decompose (from 142^{0}C for original leather up to 272^{0}C for the third pattern).

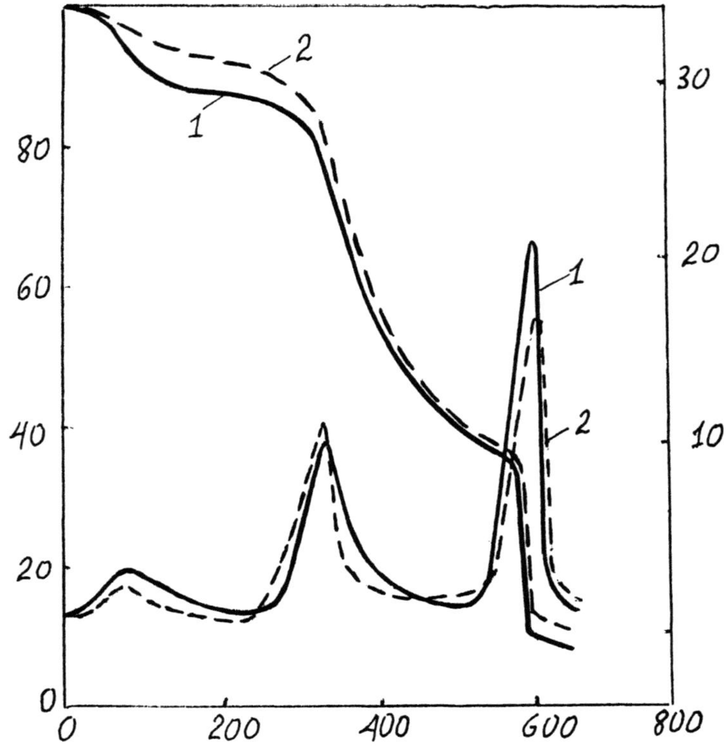

Fig.1. The results of thermogravimetric analysis:1 - original leather; 2 - the third pattern. X = temperature, 0C; Y' = mass, %; Y = oxidizing rate of carbonized residue (*V*), %/min.

The low fire-danger of the third and fourth patterns is conditioned upon their high activity in the retarding processes taking place in the gaseous phase. This activity is confirmed by a small temperature increase ($T_{max.}$) occurring during gas evolution at burning.

The data in Fig.1 give proof of the diminution of the maximal decomposition rate from 20.9 %/min. (for original leather) down to 16.5 %/min. after the introduction of melamine into the collagen dispersion containing H_3PO_4 (pattern 3). The use of the chosen retardants results not only in the alteration of the material decomposition rate, but also in the displacement of exo-effect maxima on the differential thermogravimetric

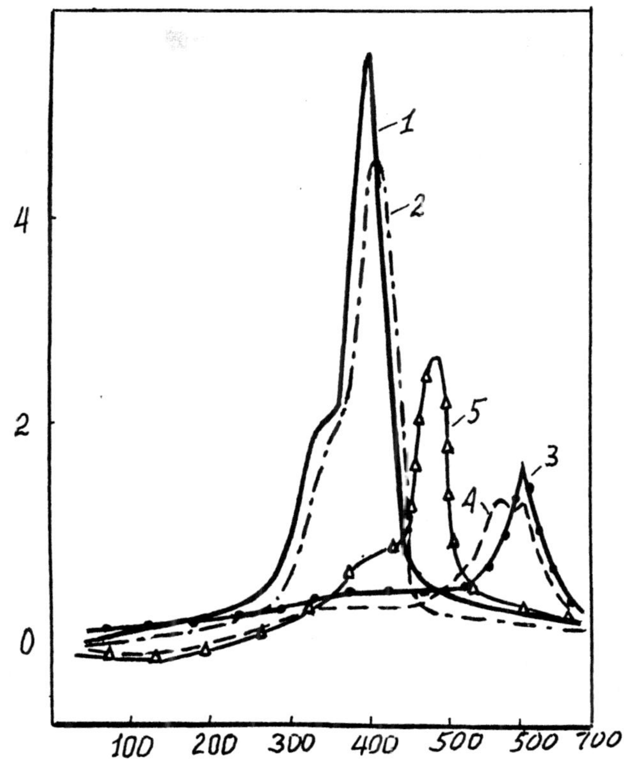

Figure 2

Fig.2. The results of differential thermal analysis of the patterns from 1 to 5.X = temperature, ^{0}C.

Y = ^{0}C/min

curves towards higher temperatures and in the lowering of the amplitude of these maxima (Fig.2).

The absence of noticeable exo-effects in the course of thermal destruction of melamine-containing compositions (patterns 3-5) must be connected with the formation of thermostable structures. These structures ensure the increase in the destruction temperature of the compositions and reduce the amount of violative products being educed which possess an inclination to be oxidized.

The data in Table 2 illustrate the increase in the oxidizing activity of formed carbonized residues. The most marked diminution of the carbonized residue's oxidizing capacity is observed for melamine-containing patterns. Thus, for the third and fourth patterns, the coke oxidizing rate goes down to 17%/min., and the quantity of coke residue at 700^0C is significantly increased.

The regularities described reveal the specificity of the flame-retarding action of melamine for collagen. This specificity consists in the suppression of oxidizing processes both in the condensed and gaseous phases. The reduction of the decomposition rate and

Table 2.*

No	Additions into collagen dispersion	Maximal oxidizing rate of carbonized residue, %/min	The amount of carbonized residue, %	
			600^0C	700^0C
1	H_3PO_4	25.1	21.9	8.4
3	H_3PO_4 + melamine	17.1	18.6	15.3
4	CH_3COOH + melamine	17.6	19.6	14.6
5	H_3PO_4 (2) + A-128	21.9	19.3	10.9
6	Original leather	20.9	26.8	7.7

* The inert atmosphere (nitrogen) was substituted by an oxidizing atmosphere (air) after 600^0C.

the lowering of heat effects (which originate from collagen oxidizing) serve as the proofs of the inhibition of the processes under discussion in the condensed phase. The oxidizing rate of the carbonized residue obtained by the thermolysis of melamine-containing material is much less than the oxidizing rate of the original material carbonized residue. In addition, the introduction of melamine into a retarding composition inhibits gaseous-phase oxidizing processes that take place at thermolysis and burning. This leads to a considerable increase in the temperature of gas evolution at combustion. Hard to ignite materials obtained from tanning industry wastes can be recommended for the design of passenger railway cars, airplane interiors, and hotels.

References

1. N.A. Khalturinskiy, T.V. Popova and A.A. Berlin, *Uspekhi Khimii,* **III** (2) (1984) 326 (in Russian).

2. M.A. Tyuganova, *Working out of the Theoretical Bases for Fire-Resistance of Spinnable Polymers and the Creation of Methods for the Production of Hard to Ignite Textile Materials* (Dissertation for the degree of Doctor of chemical science, Moscow, 1988) (in Russian).

3. V.V. Kopylov, T.A. Kachalina and L.A. Oksyenty'evich, *The Polymeric Materials with Low Combustibility* (Moscow: Khimiya, 1986) (in Russian).

4. N.S. Zubkova, T.A. Kachalina and M.A. Tyuganova, *Izv. Vuzov. Khimiya i Khim. Tekhnol.,* **35** (8) (1992) 61 (in Russian).

The Kosygin State Textile Academy,
Moscow

Part II

Chemistry, Physics and Physical Chemistry of Dyes and Textile Auxiliaries

STATE OF DYES IN SOLUTION AND THEIR ADSORPTION ON FIBRE

A.G.Zakharov, A.N.Prusov

Interrelation between state of the ionogenic dyes and adsorption processes occurring on cotton fabric is considered. State and composition of the adsorption phase at physical adsorption of dyeing substances is shown to predetermine the possibility and fixation degree of dyes by fibre. Volumetric mechanism of adsorption of the dyeing substances by fibrous material is found. A new approach to determination of the role of inorganic electrolytes in processes of physical adsorption of dyes is proposed. Mathematical correlations of quantitative interrelation between the dye activity coefficients in volumetrical and adsorption phases are deduced. Characteristic energy of adsorption is shown to characterize change in standard partial free energy at transition of 1 mole of the dye from volumetrical into adsorption phase of fibre.

Intensive development of textile materials production is accompanied by quite a number of economic and ecological problems connected in particular with increasing of effectiveness of dyes using.

It is generally known, that mass transfer processes (diffusion, sorption) of dyeing substance from liquid medium (more often aqueous medium) are influenced both of dye state in solution and fabric material properties. But most of investigations in this area have particular character or contain only qualitative and empiric conclusions about influence of different factors onto dyeing processes.

To our opinion this is caused mainly by imperfection of solution theory, especially of organic substances solutions, and weaker development of theory of adsorption from solutions as compared with theory of adsorption of gases and vapours.

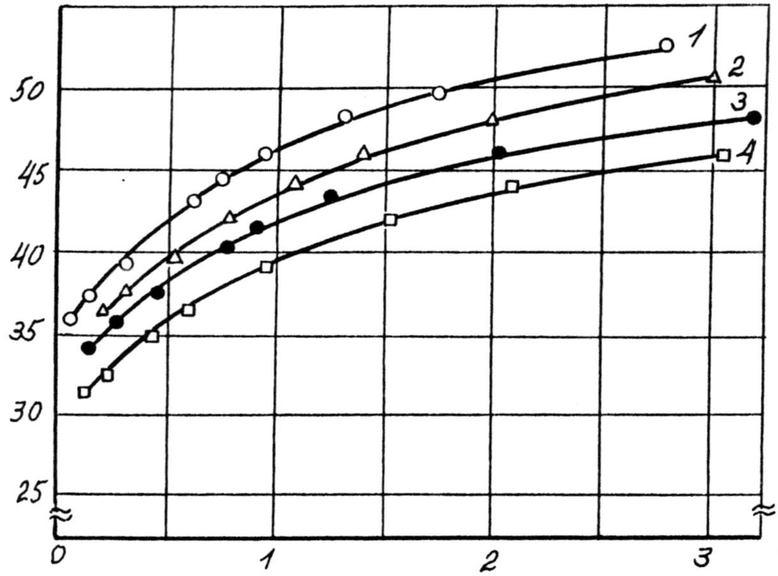

Fig.1. Enthalpy change when dissolving the dye C.I. Reactive Violet 1 in H_2O:
1 - 298 K; 2 - 308 K; 3 - 318 K; 4 - 328 K
X= m·10^2
Y= -ΔH^m_{Soln}, kJ/mole

This work presents results of investigations of thermodynamics of dyes solutions in aqueous, mixed aqueous-organic and aqueous-electrolyte solutions and thermodynamics of adsorption of the dyeing substances from these solutions. On the base of the investigations the authors suggest their own conception of quantitative interrelation between dyes state and their adsorption on fibre materials.

Thermodynamic foundation of dissolution of hydrofobic substances is closely connected with peculiarities of structure of pure and mixed solvent. When dissolving in water of the dye which differences from inorganic electrolyte by size and hydrofobity structure of water changes, it's molecules obtain more stable energetic state. Due to these changes molecules become pushed into hollows of the structure and in dyeing process - to the surface of fibre material.

Fig. 1 and 2 show concentration dependences of enthalpy change ΔH^m_{Soln} of the dye C.I. Reactive Violet 1 in aqueous and mixed solvents. In dilute solutions minimum of exothermity or maximum of endothermity is observed.

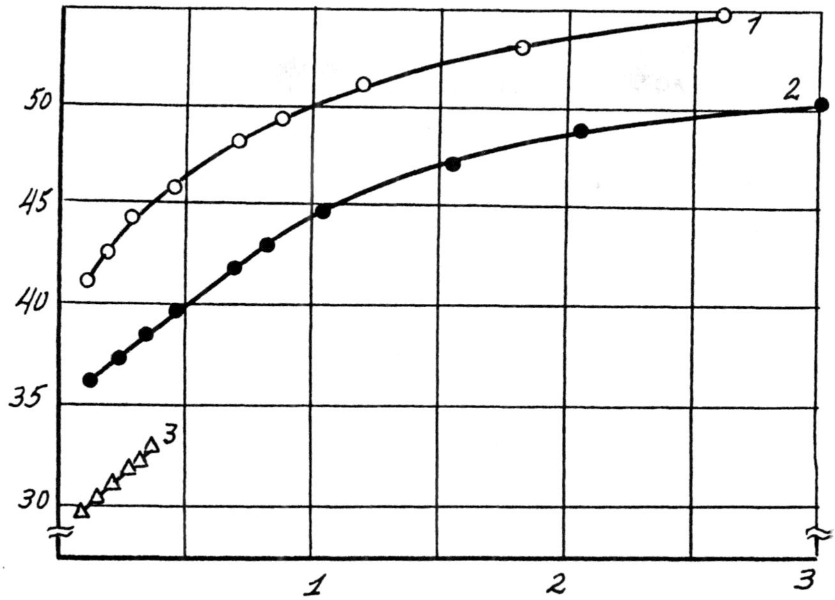

Fig.2. Enthalpy change when dissolving the dye C.I. Reactive Violet 1 in mixtures: H_2O - 0,1N NaCl (1 - 298 K; 2 - 328 K) and H_2O - 0,1 m.p. DMSO (3 - 298 K). X= m·10^2 , Y= -ΔH^m_{Soln}, kJ/mole

When concentration of the dyeing substance in dilute solutions increases exothermity of dissolution grows considerably. According to Gordon [1] structures similar to clatrat crystallohydrates have hollows surrounded with irregular polyedres. The hollows can change their structure to include inside particles of different forms giving at the same time an opportunity to all functional groups of molecules to form maximum possible number of hydrogen bonds with surrounding water molecules. However, realization of clatrat model for dyes is hardly probable, because dyes molecules are large and don't allow to form clatrat structures with large (more than 7-8) number of water molecules cycles.

Increasing of exothermity of dissolution process with increasing of dye concentration in dilute region testifies to strengthening of solvent-solvent bonds which don't require additional energy expenses. This conforms well to claster structure of aqueous solutions. According to modern view [2] taking in opinion large (as compared with solvent molecules) size of dyes ions, asymmetrical distribution and fluctuation of charge local regulating of solvent molecules close by organic ions can be observed (so called effect of dielectric saturation).

According to [3] decreasing of dielectric permeability near by ions as compared with that for solvent leads to strengthening of electrostatic interaction of the ions. In first approximation heat effect of association of the electrolyte can be presented as sum:

$$\Delta H_{ass} = \Delta H_1 + \Delta H_2 + \Delta H_3, \quad (1)$$

where ΔH_1 , ΔH_2 and ΔH_3 - enthalpy change when desolvatating of the ions (endo effect), when interacting of the ions and solvatating of associates accordingly (exo effects). For mineral salts solutions sum of ΔH_2 and ΔH_3 is smaller than ΔH_1, but for aqueous and water containing dyes solutions this sum in absolute value is larger than ΔH_1. Obviously this is connected with formation of solvato dividing dyes associates in water and water-salts solutions. The resulting heat effect of association will be determined by sign of heat effect of interaction of ions and $\Delta H_{ass} < 0$. According to the theory of intermolecule interaction energy of attraction of two identical organic ions or molecules [4] proportional to $f^2\lambda_0^3$ (λ_0 - maximum of long wave absorption band, f - oscillator force) for dye is stronger than for corresponding colourless forms. As result for able to polarization and colour ions dispersion forces can be sufficiently strong and bring appreciable contribution into final stability of the associate. On the other hand hydrofobic hydration promotes hydrofobic interaction because of aspiration of non-polar hydrocarbon fragments to get out of the aqueous surroundings decreasing additional structuration of water in hydration. Probably, exo components of mentioned processes of rapprochement of large organic ions of the dye, leading to their association, prevail upon dehydration with increase of the dye concentration.

It follows from Fig. 1 and 2 that with increasing of temperature heat effect of the dyes dissolution in water becomes less exothermic, i.e. ΔH^m_{Soln} values have positive temperature coefficients. Temperature increasing influences unfavourably upon formation of hydrate spheres destroing ready necessary for building of claster construction elements of aqueous structure. In addition positive temperature coefficients of dyes dissolution enthalpies are connected also with decreasing of exo component of the process of the association of organic ions with temperature increasing.

Thermodynamic characteristics of the process of ion solvation are presented as sum of energies of formation of the hollow in the solvent saving the solvent structure, interaction of dissolved substance with the solvent and changing of the solvent structure [5].

Forces of solvent structuration were postulated by Diamont [6]. Value of energy of the hollow formation depends upon ion size and it's hydrofobity. Value of the energy reaches maximum [1], if two hydrofobic ions simultaneously occupy the same hollow. Thus, $\Delta H^0{}_{hol}$ must be positive and decrease with increasing of the dye concentration on account of association of dissolving particles. In accordance with scale particles theory [7, 8]

$$\Delta G^0{}_{solvat} = \Delta G^0{}_{hol} + \Delta G^0{}_{int} + RT\ln(RT/V_1),$$

where $\Delta G^0{}_{hol}$ - energy, necessary for formation of one mole of hollows when brining of one mole of "solid spheres" of dissolving substance into the solvent; $\Delta G^0{}_{int}$ - energy of interaction of the dissolving substance with the solvent; $RT\ln(RT/V_1)$ - energy, characterizing changes in standard conditions.

Thermodynamic characteristics of the solvent hollow formation for reactive violet dye 4K for 298K, calculated by ratios of work [14] are represented Table 1.

Table 1

Solvent	$\Delta G^0{}_{hol}$, kJ/mole	$\Delta H^0{}_{hol}$, kJ/mole	$\Delta S^0{}_{hol}$, J/(mole·K)
Water	358.8	61.0	-999.4
DMSO, m.p.			
0.1	290.6	62.9	-764.1
0.2	258.1	63.9	-634.8
0.3	242.3	78.7	-548.8
0.5	236.6	108.5	-429.7
0.6	242.6	130.5	-376.3
0.7	253.7	159.0	-317.9
0.8	270.1	196.4	-247.4
0.9	292.4	246.7	-153.5
1.0	322.1	308.3	-46.3

Thermodynamic characteristics of the hollow formation in water, mixed and non-aqueous solvents are essentially different. Hollow formation in water for placing of large organic ions of the dye, having large hydrofobic radicals and low charge density, characterizes by considerable negative values ΔS^0_{hol} because of strong hydrogen bonds. When brining of non- aqueous solvent into water energy expenses ΔH^0_{hol} are increasing and values ΔS^0_{hol} are decreasing. In pure non- aqueous solvent (DMSO) the hollow formation is determined basically by enthalpy factor and characterizes by very small values ΔS^0_{hol}. This is obviously one of the main reasons of low solubility of the dye.

In connection with absence of proper solvation characteristics of dyes it is useful to discuss solvent influence onto properties of such solutions on the base of thermodynamic characteristics of transfer. Standard enthalpyes of transfer (ΔH^0_{trans}, kJ/mole) of individual ions and molecules of the dye C.I. Reactive Violet 1 from water into mixed solvents for 298K are represented in Table 2.

Table 2

Molecule and ion of dye	ΔH^0_{trans}, kJ/mole, for DMSO in water, m.p.								
	0.1	0.2	0.3	0.5	0.6	0.7	0.8	0.9	1.0
$R(SO_3)_3Na_3$	6.23	0.12	-7.20	-31.7	-42.9	-62.8	-72.1	-78.4	-101.7
$R(SO_3)_3^{-3}$	-3.97	6.42	21.9	34.9	34.3	18.8	12.6	8.93	-11.4

Data of Table 2 indicate of increasing of anion solvation and decreasing of the dye molecule solvation in 0.1 m.p. DMSO as compared with water. As it is known [9-11] small additions of DMSO and DMFA to water lead to substantial changes of water structure due to collective orientation of water molecules around each molecule of non-aqueous solvent. Therefore if suggest that mixed solvents form claster structures (stable in hydrofobic substances presence) dissolving of dyes with large hydrofobic radicals must lead to stabilization of the mixed solvent. Gain in energy of interaction of organic ion with the solvent is hardly probable because DMSO and DMFA considerably better solvate cations [9-11]. So considerable expense of energy in process of solvent hollow formation is presumably compensated with energy gain during reorganization the solvent structure. Further increase of DMSO concentration destroys elements of claster structure and interaction of organic ion of the dye with the mixed solvent decreases. In contrast to ΔH^0_{trans} of dye anion ΔH^0_{trans} of molecule are positive when additions of non-aqueous solvent are small, but negative when DMSO concentration increases. This is connected with changes in energy of interaction of sodium ions with mixed solvent.

Thus formation of dyes solutions proceeds with considerable and possibly determining change of entropy component of Gibbs energy.

Parallel with theoretic interest modern practice also demands investigations of dyes adsorption from solutions. Knowledge of factors which influence onto adsorption equilibrium processes, kinetics and thermodynamics of adsorption and their theoretical foundation would give an opportunity to optimize and regulate dyeing processes.

Principle difference of adsorption from solutions from adsorption of gases and vapours is that the first always has eject character and proceeds by redistribution of the solution components at the frontier of phase division but not because of gradual increase of substance concentration near adsorbent surface. Energetic characteristics, getting on the base of isotherms of adsorption from solutions, also characterise only difference between energy of adsorption interaction of molecules of the solution components with the adsorbent and energy of interaction of organic substance molecule with the solvent molecules but not pure energy of interaction of dissolved substance with the adsorbent. The energy of interaction of organic substance molecule with the solvent molecules is connected with the structure of the solvent and it's changes when organic substance dissolving, introduction of inorganic electrolyte, temperature changing, etc. Thus, total decrease of Gibbs free energy during adsorption of organic molecules from solution (ΔG_{ads}) will be determined basically by difference of values:

$$-\Delta G_{ads} = (-\Delta G_a) - (-\Delta G_s) - (-\Delta G_{sb}), \qquad (2)$$

where ΔG_a, ΔG_s and ΔG_{sb} -alterations of Gibbs energy of the system as result of interaction of the substance with the adsorbent, of the solvent with the adsorbent and of the organic molecules with the solvent correspondingly.

Complication in interpretation of isotherms of dyes adsorption on fibre is connected firstly with establishment of mechanism of the adsorption process. Most frequently for description of adsorption isotherms of dyeing substances by fibre materials equations of Langmur and Fraidlich are used [12, 13]. But all theories were used for study of adsorption from dyes solutions without taking account of structure of fibre material. Most of pores of cellulose fibre are less than 3 nm in diametre, so cellulose material may be considered as hard adsorbent with developed micropores structure. So idea about mono- or polymolecular structure of the adsorption layer is not always well-founded for fibre materials. In this case for description of adsorption equilibriums theory of volumetrical filling of micropores (TVFM) may be used. According to this theory fixation of dissolved substance under influence of aggregate field of all atoms of solid body limiting cavity of field takes place not only in region of inner surface of pores but also in volume of inside pores space. TVFM determines dependence between degree of filling of volume of adsorption space and differential molar work of adsorption. Equation of TVFM for adsorption from solutions is [14-16]:

$$\ln C_f = \ln C_\infty - (\frac{RT}{E})^n (\ln \frac{C_s}{C_i})^n, \tag{3}$$

where E- characteristic energy of adsorption, C_S - solubility of dye, C_i and C_f - equilibrium concentrations of dye in volumetrical and adsorption phases accordingly, C_∞ - maximum concentration of dye in adsorption phase, n- whole number, mainly 1,2,3.

On base of equation (3):

$$C_f = C_\infty (\frac{C_i}{C_s})^{(RT/E)^n (\ln C_s/C_i)^{n-1}} \tag{4}$$

When n=1:

$$C_f = \frac{C_\infty}{C_s^{RT/E}} C_i^{RT/E} = KC_i^b \ ,$$

where K and b are constants.

This equation is similar to equation of adsorption of Fraidlich ($C_f = aC_i^n$, where a and n are constants), in which degree index n is equal RT/E. Thus equation of Fraidlich is particular case of general equation of isotherm of adsorption in TVFM.

To use equation (2) it is necessary to know values of dye concentrations in saturated solution. A number of ionic dyeing substances are colloid electrolytes [17]. It is well known that ionic dyes have tendency to form associates. But as distinct from associates of other ionic compounds (as surfactants) there is no detailed information about structure of dyes associates. On the whole ionic fragments of ionic dyes are better hydrated with water than hydrocarbon chains of surfactants. These factors stipulate week surface-active properties of ionogenic dyes and limit the opportunity of formation of micellae of dyes without electrolytes if they don't contain large hydrofobic fragments or lateral alkyl groups. Apparently quantity and position of sulphur groups relative to large axis of symmetry of molecule play role in cases of reactive, acid and direct dyes. So if sulphur groups are at the same side of large axis of symmetry of molecule direct contact of hydrofobic fragments of the dye is largely probable. Introduction of inorganic electrolytes to dyebath increases interralation between molecules of dyeing substance and this facilitates process of micellae formation. Therefore it is necessary to take account of possibility of micellae formation to determine solubility of dyes. If micellae form critical concentration of micellae formation must be determined.

Fig. 3 and 4 show isotherms of adsorption of the dye C.I. Reactive Violet 1 and the dye Reactive Red 45 on cotton fibre from aqueous, water-salt and water-DMSO solutions. Solubility of the dyes increases when temperature increases. Combined analysis of solubility and heat effects of dissolution (ΔH^{m}_{Soln} becomes more endothermic when temperature increases) using Gibbs-Gelmgols equation ($\Delta G=\Delta H-T\Delta S$) shows that enthalpy component opposes process of dissolution of the dye when temperature increases and solubility increases owing to entropy factor. So long as adsorption is exothermal process temperature increasing must cause decreasing of value of adsorption. But it is necessary to take account of temperature influence upon solubility of the substance, i.e. upon value of it's chemical potential in solution. Superposition of these two factors determines summary influence of temperature upon adsorption equilibrium. In considered case both factors promote decrease of adsorption of the dye by cellulose material with increase of temperature.

Adsorption of the dyeing substance increases with increase of it's concentration in solution and when introducing inorganic electrolyte into dyebath. Maximum exhaust of the dyebath is observed in region of low concentrations of the dyeing substance. Obviously change of water structure caused by inclusion of molecules of dissolved substance can influence appreciably upon energy of interaction between solution components and owing to this influence upon equilibrium distribution of the dye between adsorbent and solvent. First of all in dilute region of concentrations on account of local regulating of solvent molecules nearby organic ions (dielectric permeability decreases) interaction between diverse charged ions increases but not interaction of dissolved substance with solvent. As a result sharp decrease of coefficients of activity in dilute region of concentrations must be observed, that is confirmed with experimental facts [18]. Further increase of concentration of dissolved substance leads to increase of interaction between same charged dye anions (association occurs) and as effect according to equation (1) energy of interaction of associate of dyeing substance with solvent may increase. According to equation (2) this must lead to deceleration of exhaust of the dyebath if dissolved substance presents in it in considerable concentrations. It is shown [12], that introduction of inorganic electrolyte into aqueous solution of dyeing substance influences firstly on double electric layer of fibre material and thus promote increase of exhaust the dissolved substance. However as stated above inorganic electrolyte also decreases energy of interaction of water both with the dye and cellulose fibre. That in accordance with (2) must promote adsorption process. As a result transfer of the dye from volumetrical into adsorption phase may influence on differential heats of adsorption, because weakening of interaction of water with the dye and fibre must lead to increase of degree of their dehydration during adsorption. Changes of differential molar enthalpy of adsorption of the dye C.I. Reactive Violet 1 on cellulose fibre from water and 0.1N solution of sodium chloride (Fig. 5) confirm supposition stated above. Data of Fig. 5 testify also to substantial role of entropy factor in process of adsorption of dye with fibre material.

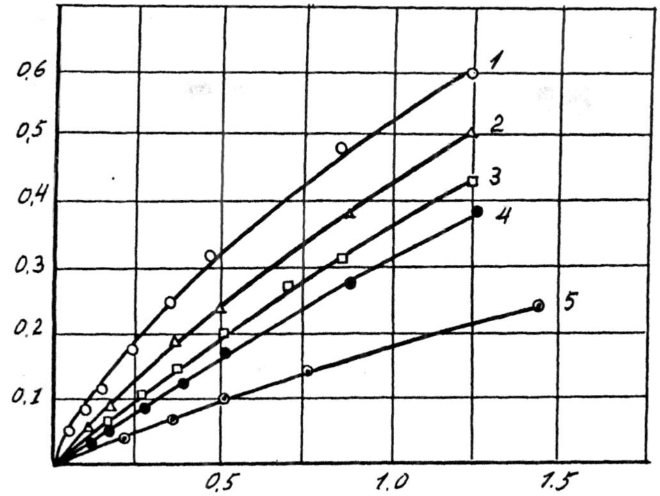

Fig. 3. The adsorption isotherms of the dye C.I. Reactive Violet 1 (1 - 298K, Cs=0.129 mole/l, $C_\infty=3.16 \cdot 10^{-2}$ mole/kg; 2 - 308K, Cs=0.141, $C_\infty=3.05 \cdot 10^{-2}$; 3 - 318K, Cs=0.15, $C_\infty=3.17 \cdot 10^{-2}$; 4 - 328K, Cs=0.159, $C_\infty=3.16 \cdot 10^{-2}$) and the dye Reactive Red 45 (5 - 298K, Cs=0.153, $C_\infty=1.67 \cdot 10^{-2}$) by cotton fibre from aqueous solutions. $X= C_i \cdot 10^2$, $Y= C_f \cdot 10^2$

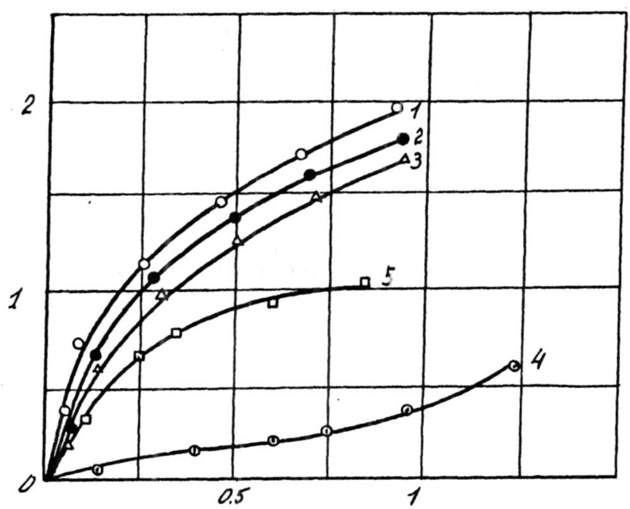

Fig. 4. The adsorption isotherms of the dye C.I. Reactive Violet 1 (1 - 298K, Cs=0.125 mole/l, $C_\infty=3.13 \cdot 10^{-2}$ mole/kg; 2 - 308K, Cs=0.134, $C_\infty=2.97 \cdot 10^{-2}$; 3 - 318K, Cs=0.145, $C_\infty=3.04 \cdot 10^{-2}$; 4 - 298K, Cs=0.021) and the dye Reactive Red (5 - 298K, Cs=0.14, $C_\infty=1.64 \cdot 10^{-2}$) by cotton fibre from mixtures H_2O- 0.1N NaCl (1,2,3,5) and H_2O-0.1 m.p. DMSO (4).

$X= C_i \cdot 10^2$ $Y= C_f \cdot 10^2$

Attracts attention the fact that addition of DMSO to aqueous solutions of the dye changes form of isotherms of adsorption. In this case they have characteristic salient to axis of abscissas form. This testifies to week interaction of the dye with cellulose fibre and to increase of energy of interaction of mixed solvent with fibre material. Obviously, it is connected with change of energy of inter particle interactions not only in volumetrical but also in adsorption phases. Possibly, in the last composition of mixed solvent will differ from the same in volumetrical phase, so as solvating ability of water and DMSO in regard to cotton fibre is different (DMSO solvates cellulose material better). Received data of solubility of series of reactive dyes in water-DMSO solutions show sharp decrease of their solubility as compared with water. Therefore, decrease of solubility of dyeing substance may be criterium of increase of adsorption not in all cases.

Treatment of experimental data of Fig. 3 and 4 showed that when going from aqueous to water-salt solutions value of n [equation (3)] changes from 1 to 2. At the first view this fact is unexpected, because this change is connected first of all with increase of adsorption potential, which is determined by size of pores of the adsorbent. According to [19] phenomenon of physical adsorption on micropore adsorbents is analogous thermodynamically with phenomena of volumetrical dissolution. Thermodynamical base of adsorption in this case is not Gibbs equation but Gibbs-Dugem equation. Chemical potential of adsorbent changes during adsorption of ions (molecules) similar to change of chemical potential in process of solution formation. So it may be suggested that introduction of inorganic electrolyte into dye solution is accompanied by change of chemical potential not only of dissolved substance but of adsorbent also. This leads to increase of equilibrium adsorption and change of adsorption potential.

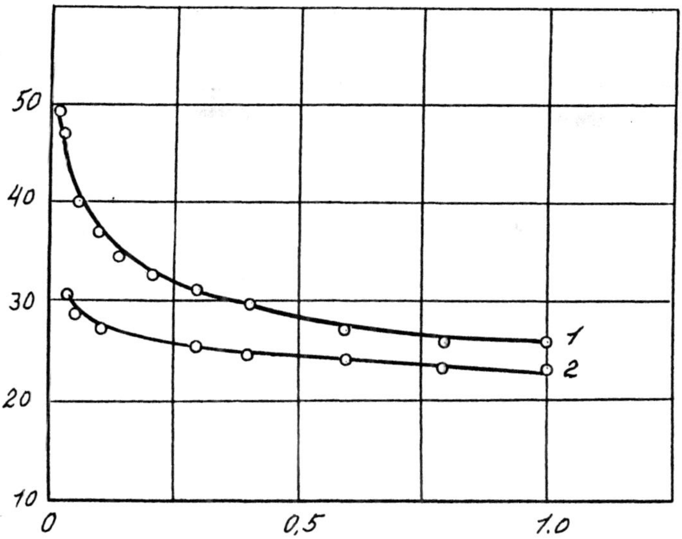

Fig. 5. The Differential heat of adsorption of the dye C.I. Reactive Violet 1 to cotton fibre at 298K: 1 - from H_2O; 2 - from 1.2N NaCl solution.

$X=C_f \cdot 10^3$ $Y=-\Delta\overline{H}_{ads}$, kJ/mole

Accepting thesis that transition from big-pores sorbents to micro-pores sorbents is accompanied by disappearance of surface of phase division and sorbent-sorbat system becomes in a sense monophase system we'll consider that walls of pores of cellulose fibre become one of the components of the solvent in regard to the dye. In this case not individual centres on the surface but each atom of adsorbent surface take part in interaction with atoms of adsorbed molecules. Therefore in limits of field of action of adsorption forces molecules save their mobility and in adsorbent phase act so as in solution of corresponding composition, differ only with less coefficients of diffusion and character of interaction with each other because of influence of the adsorption field.

In this connection we'll consider the adsorption equilibrium as equilibrium of the dye solutions in volumetrical and adsorption phases.

Motive force of adsorption is difference of chemical potentials of the dyeing substance in volumetrical and adsorption phases. Condition of the dissolved substance in hypothetical ideal solution ($\gamma_i=1$) with unity concentration assume to be it's standard condition in real solution (both in volumetrical and adsorption phases) [20].

Change of the chemical potential (partial molar Gibbs energy) of the dissolved dye in process of formation of real solution in volumetrical and adsorption phases may be written as

$$\Delta\overline{G}_f = \Delta\overline{G}_f^0 + RT\ln a_f \qquad (5)$$

and

$$\Delta\overline{G}_i = \Delta\overline{G}_i^0 + RT\ln a_i \ ,$$

where a_f and a_i -activity of the dye in volumetrical and adsorption phases correspondingly.

We'll have in mind, that

$$\Delta\overline{G}_f^0 = -RT\ln a$$

and

$$\Delta\overline{G}_i^0 = -RT\ln a_S \ .$$

If two phases are in equilibrium, than

$$\Delta\overline{G}_f = \Delta\overline{G}_i$$

and

$$-(\Delta\overline{G}_f^0 - \Delta\overline{G}_i^0) = -\Delta G_{ads}^0 = RT(\ln\frac{C_{V_f}}{C_i} + \ln\frac{\gamma_f}{\gamma_i}) = RT(\ln\frac{C_{V_¥}}{C_S} + \ln\frac{\gamma_¥}{\gamma_S}) \qquad (6),$$

where $C_{V_¥}$ and C_{V_f} -dye concentrations in fibre, mole/dm^3. They are connected with C_∞ and C_i (mole/kg) by expressions:

$$C_{V_\infty} = \frac{C_\infty}{V_\infty} \text{ and } C_{V_f} = \frac{C_f}{V_¥} \ ,$$

where V_∞ - maximum accessible for dye volume of the fibre material.

On the other hand when chemical potentials of the dye in volumetrical and adsorption phases are equal it may be written:

$$-RT \ln a_\infty + RT \ln a_i = -RT \ln a_S + RT \ln a_i \qquad (7).$$

Transforming equation (7), we'll receive

$$\ln\frac{\gamma_{¥}\gamma_{i}}{\gamma_{f}\gamma_{s}} = \ln\frac{C_{s}C_{f}}{C_{¥}C_{i}} \qquad (8).$$

Combined solution of equations (5) and (8) leads to equation

$$\ln\frac{\gamma_{¥}\gamma_{i}}{\gamma_{f}\gamma_{s}} = (\ln C_{i}/C_{S})[1-(\frac{RT}{E})^{r}(\ln C_{S}/C_{i})^{n-1}).$$

To find $-\Delta G^{0}_{ads}$, which characterises standard affinity of the dye with the fibre, it is necessary to know maximum volume of pores V_∞ and relation ln γ_∞/γ_s or ln γ_f/γ_i. Other values may be found easily from adsorption isotherm by equation (3).

To find maximum adsorption volume equation

$$V_\infty = C_\infty v^* \qquad (9),$$

where v^* - molar volume of the dye, is used.

Necessary for calculations values of molar volumes of dyes may be found from their density in solid state. This value may be calculated also as product of area of surface of projection of absorbed molecule and it's size, normal to plane of projection, i.e. so called thickness of the molecule. To determine them Stuart-Brigleb models of the dyes were assembled. When adsorption of ions of dyes takes place on charged surface of cotton fibre orientation of planes of their benzene rings parallel to adsorbent plane is not always possible. In case of adsorption of the dye C.I. Reactive Violet 1 having three sulphur groups (SO_3^-) at the same side of large axis of symmetry of the molecule orientation of ions of the dye by "edges" of aromatic rings to turn sulphur groups deep into solvent due to strong ion-dipole interaction of sulphur groups with solvent molecules becomes more probable. In case of adsorption of the dye Reactive Red 45 such type of orientation is energetically disadvantageous because three sulphur groups dispose at both sides of large axis of symmetry of the molecule, so only some sulphur groups will turn deep into solvent. Therefore choice of landing ground for ionogenic dyes has to some extent arbitrary character. Areas of projection ω and their molar volumes list in table 3.

Table 3.

Dye	ω, A^2/molec.	v^*, dm^3/mole (calculated as M/ρ)	v^*, dm^3/mole
C.I. Reactive Violet 1	121	0.50	0.53
C.I. Reactive Red 45	200	0.50	0.54

Calculated with both methods molar volumes of the dyes are in good accord with each other. Received maximum adsorption volumes V_∞ of dyes C.I. Reactive Violet 1 and C.I. Reactive Red 45 are equal to $1.7\cdot 10^{-2}$ and $9{,}0\cdot 10^{-3}$ dm^3/kg correspondingly (considerably less than listed in works [12, 21] - 0.22-0.3 dm^3/kg). This is because in process of adsorption from solutions we deal with two components (molecules of dye and water) as minimum, both components presenting in fibre pores. Their ratio is determined by interaction between dye ions and water molecules both in volumetrical and adsorption phases. Therefore there is no reasons to consider that adsorption of the dye is accompanied with total supplanting of the solvent. So when saturating cotton fibre with the dye adsorption volume remains two-component. Thus using of equation (9) to determine maximum adsorption volume of pores of adsorbent is possible only in cases when energy of interaction of the dye with the solvent is smaller than energy of interaction between the dye molecules and macromolecules of fibre material and difference of adsorption energies of both components is considerable.

If sieve effect is absent, i.e. part of micro pores becomes inaccessible for the dye because of steric causes volume of pores of swelled in water cotton fibre is equal to 0.22 l/kg [22, 23]. In dependence $\ln(\gamma_\infty\gamma_i/\gamma_f\gamma_S)=f(C_i)$ [see equation (8)] during adsorption of dyes C.I. Reactive Violet 1 and C.I. Reactive Red 45 from water-salt solution point of bend was found. In this point it may be $\ln(C_S C_f/C_\infty C_i) = \ln(\gamma_\infty/\gamma_S)$. By substitution of found value $\ln(\gamma_\infty/\gamma_S)$ (in case of the dye C.I. Reactive Violet 1) and V_∞ =0.22 l/kg into equation (6) value of $\Delta\overline{G}^{\,0}_{ads}$ was found. It proved to be similar to received in experiment value of characteristic energy of adsorption E (table 4). Obviously for process of adsorption of dyeing substances on cellulose material (in conditions of multicomponent filling of micro pores) value of characteristic energy of adsorption characterises change of standard partial free energy during transfer of one mole of the dye from volumetrical into adsorption phase. In case of the dye C.I. Reactive Red 45 presence of sieve effect was established earlier [24]. Because of this by equation (6) using equality $-\Delta\overline{G}^{0}_{ads} = E$ and found value of $\ln(\gamma_\infty/\gamma_S)$ maximum adsorption volume V_∞ accessible for the dyeing substance was calculated (see table 4). Received results testify that it is necessary to determine accessible for dye volume of fibre material from experiment with that dye and this volume can't be the same for all dyeing substances.

Table 4.

Dye	Water, solution at 298K	V_∞, l/kg	C_i, mole/l at C_f=0.368C_∞	$\ln(\gamma_\infty/\gamma_S)$ [equation (6)]	$\ln(\gamma_\infty/\gamma_S)$ [equation (8)]	E, J/mole	$-\Delta\overline{G}^{0}_{ads}$ J/mole
C.I. Reactive Violet 1	Water	0.22	$3.184 \cdot 10^{-2}$	1.29	–	3465	-
	0.1N NaCl solution	0.22	$2.69 \cdot 10^{-3}$	3.71	3.69	9504	9458
C.I. Reactive Red 45	Water	0.14	$4.56 \cdot 10^{-2}$	1.46	–	3000	-
	0.1N NaCl solution	0.14	$2.15 \cdot 10^{-3}$	–	4.36	1034	-

Difference in value of E for dyes when adsorption from water and water-salt solution (see table 4) is connected with the fact that in the last case we have E at present ionic force (which is conditioned by presence of inorganic electrolyte). Adsorption isotherms of the dye C.I. Reactive Violet 1 from water-salt solutions to cotton fibre are presented in Fig. 6.

Treatment of the experimental data by equation (3) showed that at concentrations of sodium chloride $3.42 \cdot 10^{-3}$; $8.55 \cdot 10^{-3}$;$1.71 \cdot 10^{-2}$ mole/l values of E are different (Fig. 7, line 1) at n=1, but at concentrations $3.42 \cdot 10^{-2}$; $6.84 \cdot 10^{-2}$; 0.1 mole/l not only value of E changes

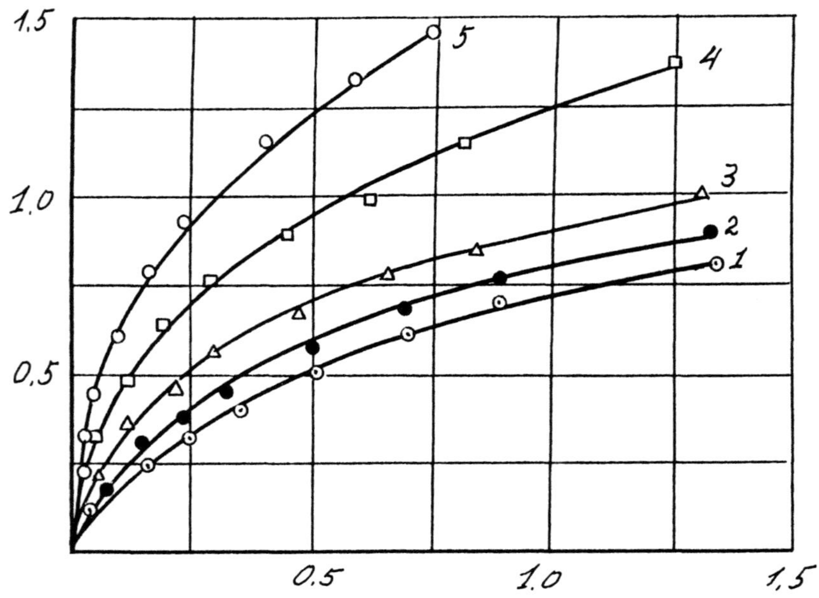

Fig. 6. Adsorption isotherms of the dye C.I. Reactive Violet 1 from aqueous solutions of sodium chloride: 1 - $3.42 \cdot 10^{-3}$N; 2 - $8.55 \cdot 10^{-3}$N; 3 - $1.71 \cdot 10^{-2}$N; 4 - $3.42 \cdot 10^{-2}$N; 5 - $6.84 \cdot 10^{-2}$N.
$X = C_i \cdot 10^2$ $Y = C_f \cdot 10^2$

(Fig. 7, line 2), but also degree index n [see equation (3)] changes (n=1.2; 1.6 and2 correspondingly). Extrapolation of received rectilinear dependences 1 and 2 (see Fig. 7) to zero ionic force gives values of E=3575 and 3461 J/mole correspondingly. Found values are similar to E for aqueous solution (3465 J/mole).

In case when $E = -\Delta\overline{G}^0_{ads}$ using equation (3) it may be written:

$$\ln C_f = \ln C - \left(\frac{RT}{-\Delta\overline{G}^0_{ads}}\right)^n \left(\ln\frac{C_s}{C_i}\right)^n . \quad (10)$$

Independently from value n at $C_f = 0.368\ C_\infty$ by transformation of equation (10) it may be written:

$$E = -\Delta\overline{G}^0_{ads} = RT\ln C_S / C_i = RT\ln a\ / a_S = RT\ln a_f / a_i . \quad (11)$$

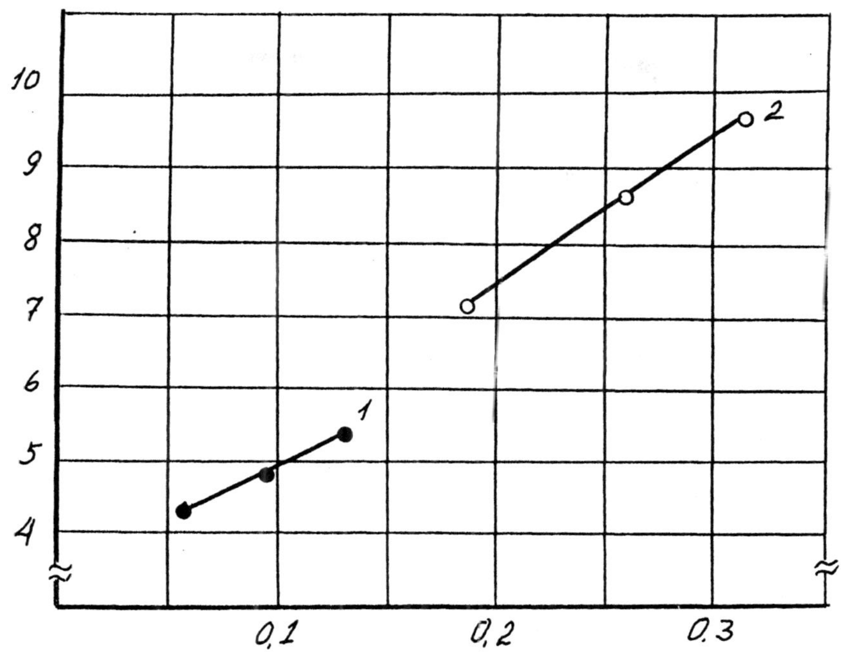

Fig. 7. Dependence E from ionity of solution at 298K for the dye C.I. Reactive Violet 1 . X= $\sqrt{I}$ Y= E,kJ/mole

From equation (11) following relations may be received:

$$\ln \gamma_f / \gamma_i = \ln(C_s V / C_f);$$

$$\ln \gamma \, \gamma_S = \ln(C_S^2 V / C_i C).$$

Considered material corroborates thermodynamical theory and shows, that when investigating of adsorption equilibrium activities of the dye (both in volumetrical and adsorption phases) but not of ion of the dyeing substance are of great importance.

Probably, choice of concentration limit of adding inorganic electrolyte, leading to significant increase of dye absorption may be carried out by degree index n in equation (3). The minimum concentration of the electrolyte at which n passes from 1 to2 will be optimum for mass transfer process of the dyeing substance into cellulose fibre. Further increase of the electrolyte concentration increases density of dyes packing limiting mobility and accessibility of active centre of the dyeing substance for hydroxyl groups of the fibre.

And this may lead to decrease of quantity of covalent fixed substance and deterioration of fastness characteristics of dyeing. Obviously, increase of absorption of reactive dye may be expected in solutions with greater ΔH^{m}_{Soln}. The same effect may be obtained by introducing of small amounts of non-aqueous solvent with lesser solvatating ability with respect both to dye and fibre. Mixed solvent must has lesser solvatating ability to cation and anion of the dyeing substance as compared with water.

For direct dyes with greater affinity with cellulose addition of even small amounts of inorganic electrolyte may lead to adsorption of the dyeing substance close by pores mouth and part of volume of the fibre material will become inaccessible. Besides that in this conditions equilibrium state will be reached slowly (Rate of dyeing will decrease).

Presented material shows that development of physico-chemical bases of dyeing is impossible without fundamental investigations of thermodynamics and energetics of the processes occurring during formation of dyeing solutions and mass transfer processes of dyes to fibre. Proposed approach (to consider system cellulose material - adsorped dye as single phase system as adsorption solution) allows to use thermodynamics of solutions for processes of dyes adsorption on polymer substrate.

References

1. G.Gordon, Organic Chemistry of the Electrolyte Solutions (Moscow: Mir,1979) 712 (in Russian)

2. S.G.Entelis and R.P.Tiger, Kinetics of the Liquid Phase Reactions. Quantitative Calculation of Influence of Surroundings (Moscow: Khimiya,1973) 416 (in Russian)

3. Yu.M Kessler, Non-coulomb Effects in Thermodynamics of the Strong Elecrolyte Solutions (Dissertation for the degree of Doctor ofChemical Sciences. Ivanovo,1969) 327 (in Russian)

4. V.I.Yujakov, Uspechi Khimii, 48 (1979) 2007 (in Russian)

5. O.E.Kiselev and G.A.Martinov, in Collected Article: Theoretical Methods of Description of the Solution Properties (Ivanovo,1987) 29 (in Russian)

6. K.S.Pitzer, Accounts Chem.Res, 10 (10) (1977) 371

7. H.Reiss, H.L.Frisch and E.Helfand, J.Chem.Phys. 32 (1) (1960) 119

8. L.P.Safonova, A.N.Kinchin and A.M.Kolker, J. Phys.Chem., 63 (11) (1989) 2919 (in Russian)

9. R.Fuchs, C.P.Hagan and R.E.Rodewald, J. Phys.Chem., 78 (15) (1974) 1509

10. M.Booij and G.Somsen, Electrochem.Acta, 28 (12) (1983) 1883

11. B.E.Geller, J. Phys.Chem., 35 (5) (1961) 1105 (in Russian)

12. T Vikkerstaff, Physical Chemistry of Dyeing, (Moscow: Gizlegprom,1956) 574 (in Russian)

13. M.A.Kulikova and G.E.Kritchevsky, Izvestiya Vuzov. Technotext.Prom. (4) (1977) 59 (in Russian)

14. M.M.Dubinin, in Collected Article: The Basical Problems of Theory of the Physical Adsorption (Moscow: Nauka,1970) 251 (in Russian)

15. M.M.Dubinin and V.A.Astachov, Izvestiya AN SSSR. Seriya Khimiya (1) (1971) 5 (in Russian)

16. A.M.Koganovsky and T.M.Levchenko, J. Phys.Chem., 46 (7) (1972) 1789 (in Russian)

17. K Mitchel, Micelloformation, solubilisation and micro emulsions (Moscow: Mir,1980) 499 (in Russian)

18 A.G.Zakharov, Thermodynamics of Solvation and Dyeing in the Ionogenic Dyes Solutions (Dissertation for the degree of Doctor of Chemical Sciences. Ivanovo,1990) 290 (in Russian)

19. B.P.Bering and V.V.Serpinsky, in Collected Article: Adsorption and Porosity (Moscow: Nauka,1976) 231 (in Russian)

20. V.K.Abrosimov, in Collected Article: Modern Problems of the Solution Chemistry (Moscow: Nauka,1986) 97 (in Russian)

21. N.I.Klenkova, Structure and the Reaction Ability of Cellulose (Leningrad: Nauka,1976) 368 (in Russian)

22. M.Mitsuishi et al, J.Soc.Fiber Sci and Technol. 40 (1) (1984) 107

23. F.G.Lupashku, A.V.Mamtchenko and V.N.Ropot, Kolloidny Jurnal 46 (2) (1984) 364 (in Russian)

24. A.N.Prusov et al, DAN SSSR 310 (3) (1990) 646 (in Russian)

The Institute of Chemistry of Non-Aqueous Solutions

of the Russian Academy of Sciences, Ivanovo

THE INFLUENCE OF HIGH-SPEED BLOW-PULSE LOADING ON THE SUBLIMATION OF DISPERSE DYES

V.A.Padokhin, G.A.Zueva, V.N.Blinitchev

By means of the thermo-gravimetric method, it has been established that the high-speed blow-pulse loading of disperse dyes (with C.I. Disperse Red 4 and C.I. Disperse Red 60 serving as the examples) causes a change in the activation energy of dye sublimation. Empirical relations have been developed which describe the dependence of the activation energies of sublimation of the above-mentioned dyes upon the blow-pulse loading speed. Regularities have been revealed of the evolution of dye sublimation activation energies initiated by mechanical loading.

In the field of synthetic textile material coloring, the technologies connected with the application of disperse dye vapors (obtained by sublimation) are one of the most important [1, 2]. The application of disperse dye vapors for textile material dyeing opens wide possibilities for the creation of a new generation of energy saving, environmentally safe technologies in which water consumption will be completely eliminated.

At present, experimental and statistic information on heat- and mass-transfer phenomena during sublimation textile dyeing processes is being accumulated and generalized, and the means for the intensification of such processes are being created.

The aim of the present article consists in studying the correlation between the blow-pulse loading of disperse dyes and their sublimation. Two dyes, Disperse Red 4 (C.I. 60755) and Disperse Red 60 (C.I. 60756), were chosen for our investigations.

In conformity with the existing classification of disperse dyes according to their transfer rates from a paper backing to the polyester fiber [3], Disperse Red 60 and Disperse Red 4 belong respectively to those groups of dyes with mean and moderately high sublimation rates.

The blow-pulse loading of the disperse dyes under investigation was performed using a purpose-built laboratory unit (see Fig.1) which works as follows: The disperse dye which must be loaded is fed into the rotor-impulse apparatus through its loader with the help of a dosator. Moving together with the air flow, the dye particles gather speed (due to the quick gyration of the elements fixed on the rotor) and strike against the reflecting elements. After loading, the dye particles are taken away from the unit through a branch pipe by means of air flow. The alteration of the speed of dye loading was performed by changing the rotor gyration frequency. The rotor was driven by an asynchronous alternating current electric motor through pulleys and a belt transmission.

The activation energy of disperse dye sublimation before and after blow-pulse loading was defined by thermo-gravimetric analysis which was performed using the a Q-1500D apparatus.

The experimental data thus obtained were processed according to the procedures given in detail in the manuscripts [4, 5].

Fragments of the whole experimental gravimetric data on two disperse dyes before and after their blow-pulse loading are shown in Fig.2. It is clearly seen that an increase in loading speed leads to a rise in the dye sublimation rate and to the displacement of melting zones towards lower temperatures.

The activation energies of dye sublimation were calculated on the basis of the experimental thermo-gravimetric data using Langmuir and Clausius-Clapeyron equations. Graphs demonstrating the dependencies of the activation energies of the sublimation of two disperse dyes upon their loading speed are given in Fig.3. It is apparent that in the studied interval of loading speed alteration, the activation energies of dye sublimation decrease when the loading speed is increased. Such a regularity is obviously caused by the increase in the concentration of different dye structure defects during the mechanical loading of solid dye particles [6].

The analysis of the activation energy evolution (initiated by blow-pulse loading) of two disperse dyes showed that these energies relax towards their corresponding heats of sublimation (Fig.4). To all appearances, this fact can be explained by "treating" the structural defects in dye particles through time.

The mathematical processing of the obtained experimental data gave the following

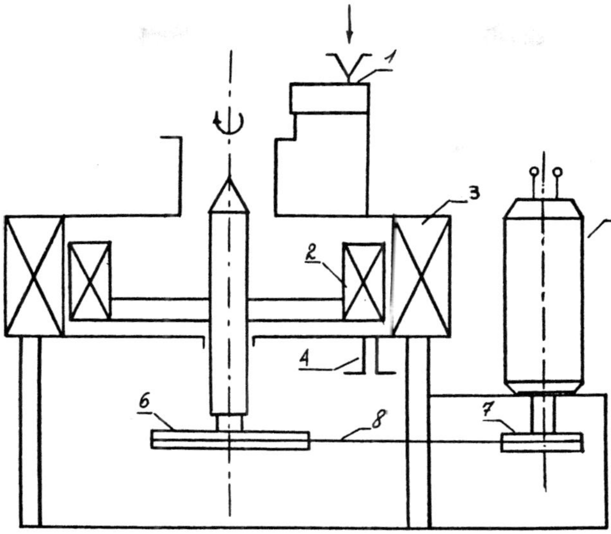

Fig.1. A general diagram of the laboratory unit used for research into the influence of dye blow-pulse loading on disperse dye sublimation: 1 - the dosator; 2 - rapidly gyrating elements; 3 - reflecting elements; 4 - the branch pipe; 5 - the asynchronous electric motor; 6, 7 - pulleys; 8 - the belt transmission.

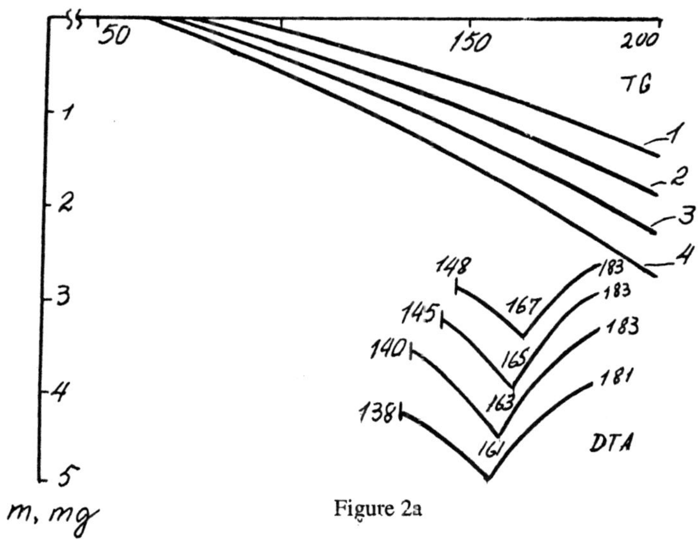

Figure 2a

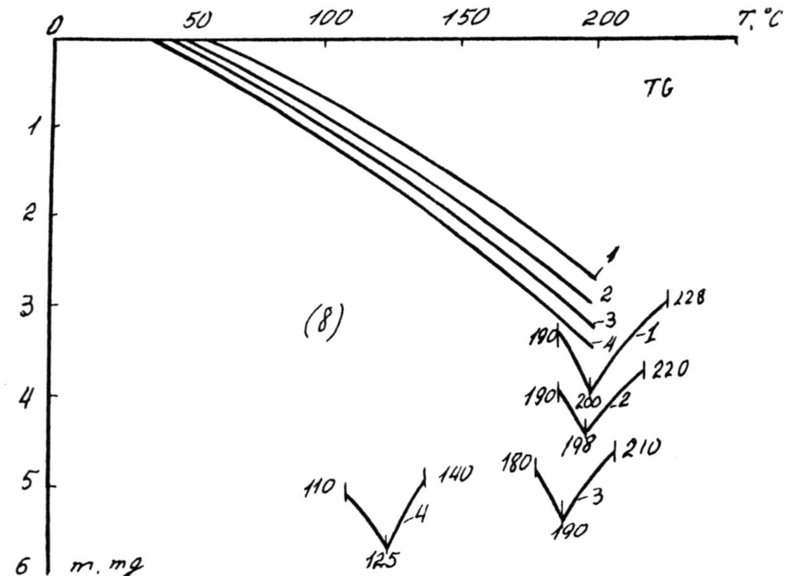

Fig.2.b

Fig.2. Fragments of thermo-gravimetric data experimentally obtained for C.I. Disperse Red 60 (a) and C.I. Disperse Red 4 (b): 1 - before blow-pulse loading; 2, 3, 4 - after blow-pulse loading at speeds of 64 m/s, 85 m/s, and 113 m/s, respectively. X = temperature, ^{0}C; Y = m, mg.

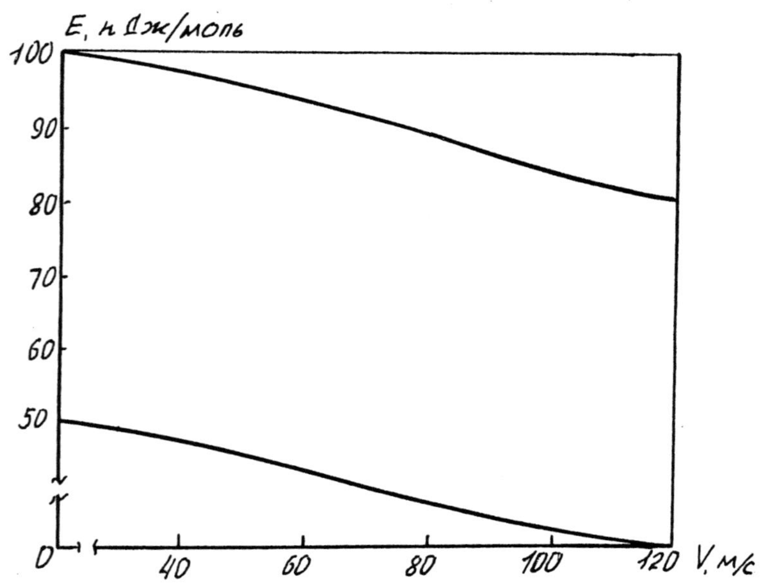

Fig.3. The dependencies of the activation energies of sublimation of C.I. Disperse Red 60 (1) and C.I. Disperse Red 4 (2) upon dye blow-pulse loading speed. X = temperature, ^{0}C; Y = m, mg.

inter-correlations between the activation energy of dye sublimation and the dye loading speed:

for C.I. Disperse Red 60:$E_{D.R.60} = 98.96 \exp(-1.75 \cdot 10^{-5} \cdot V^2)$;

for C.I. Disperse Red 4:

$E_{D.R.4} = 49.33 \exp(-1.54 \cdot 10^{-5} \cdot V^2)$.

The variations in the activation energy of sublimation of loaded disperse dyes through time were approximated by the following expression:

$$E = E_{\min.} + (E_0 - E_{\min.}) \cdot \left(1 - \exp\left(\frac{t}{t_p}\right)\right)$$

where $E_{\min.}$ = the activation energy of dye sublimation just after the loading of the dye; E_0 = dye sublimation heat; t_p = a certain constant; and t = current time.

As a result of the mathematical processing of the experimental curves in Fig.4, the dependencies of the activation energy of dye sublimation upon dye treatment time were obtained:

for C.I. Disperse Red 60:

$E_1 = 94.42 + 4.54\ (1 - \exp(-t/40428))$ at V = 64 m/s,

$E_2 = 83.76 + 15.2\ (1 - \exp(-t/20428))$ at V = 85 m/s,

$E_3 = 80.88 + 18.08\ (1 - \exp(-t/126428))$ at V = 113 m/s,

for C.I. Disperse Red 4:

$E_4 = 47.25 + 2.08\ (1 - \exp(-t/86939))$ at V = 64 m/s,

$E_5 = 49.83 + 5.5\ (1 - \exp(-t/86939))$ at V = 85 m/s,

$E_6 = 40.04 + 9.32\ (1 - \exp(-t/106939))$ at V = 113 m/s.

Thus, the experimental investigations performed resulted in the establishment of the fact that the blow-pulse loading of disperse dyes leads to the diminution of the activation energy of sublimation of C.I. Disperse Red 60 and C.I. Disperse Red 4. It is also shown that in the studied interval of dye loading speed alteration, the activation energies of dye sublimation relax towards their heats of sublimation.

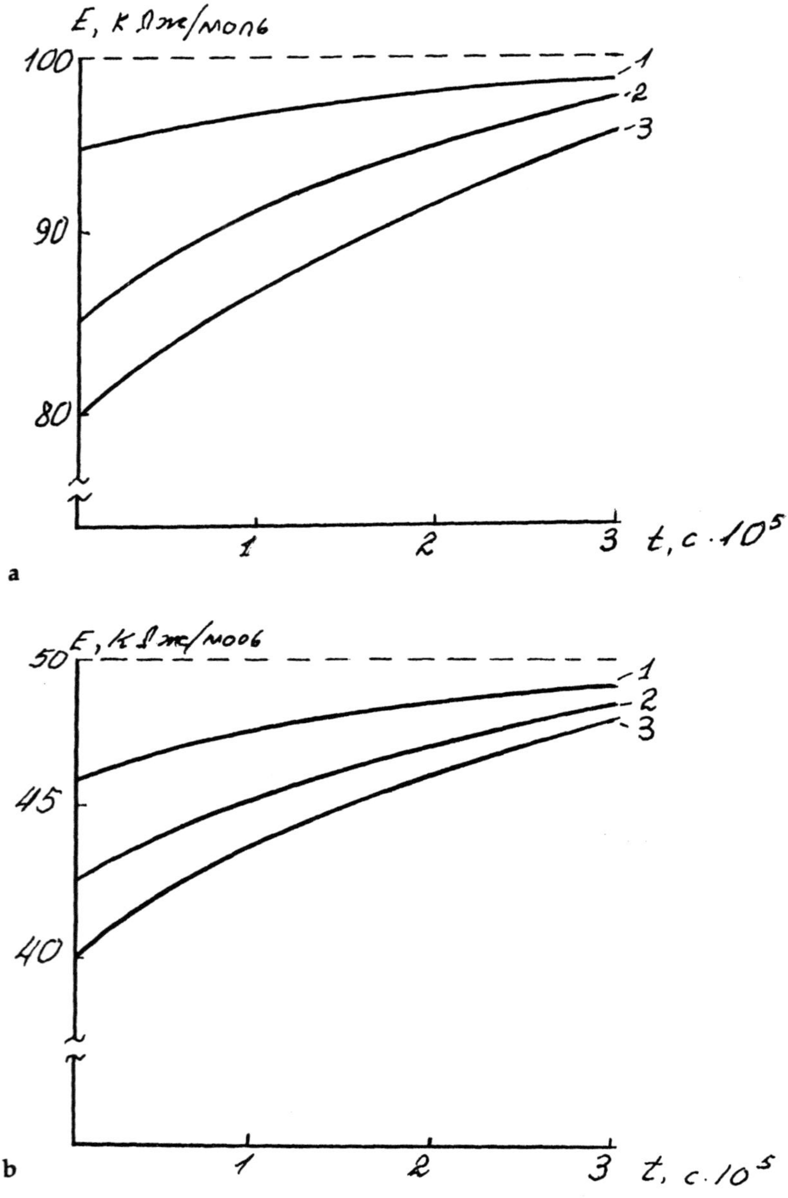

Fig.4. The variations of the activation energies of sublimation of C.I. Disperse Red 60 (a) and C.I. Disperse Red 4 (b) (initiated by blow-pulse loading) as a function of time: 1 - V = 64 m/s, 2 - V = 85 m/s, 3 - V = 113 m/s. X = V, m/s; Y = E, kJ/mol.

References

1. B.N. Mel'nikov , M.N.Kirillova and A.P. Moryganov, The Present Situation and the Possibilities for the Development of Textile Material Dyeing Processes (Moscow: Lyogkaya i Pishchevaya Promyshlennost, 1983) (in Russian).

2. B.N. Mel'nikov, A.P. Moryganov and Yu.A. Kalinnikov, The Theory and Practice of Intensive Fixation of Dyes on Textile Materials (Moscow: Legprombytizdat, 1984) (in Russian).

3 V.V. Karpov, I.V. Krutovskaya and G.N. Rodionova, in Collected articles of Leningrad Textile Institute: The Problems of Chemistry and Environment in Textile Industry (Leningrad: Publishing Agency of the Leningrad Textile Institute, 1979) 126 (in Russian).

4. W. Wendland, The Thermal Methods of Analysis (Moscow: Mir, 1978) (Russian Edition).

5. V.S. Gorshkov, V.V. Timashev and V.G. Savel'ev, The Methods of Physicochemical Analysis of Cementing Materials (Moscow: Vysshaya shkola, 1981) (in Russian).

6. N.A. Bogolyubov and S.I. Masharov, The Evaporation of the Atomically and Magnetically Ordered Crystals Containing Structural Defects (Novosibirsk: Nauka, 1989) (in Russian).

The Institute of Chemistry of Non-Aqueous Solutions
of the Russian Academy of Sciences, Ivanovo

THE SYNTHESIS AND PROPERTIES OF A NEW REDUCING AGENT: SODIUM HYDROXY-ISO-PROPYLSULFINATE

E.L.Kadykova, S.M.Gubina

The peculiarities of the synthesis of sodium hydroxy-iso-propylsulfinate and the data on the reaction ability of this substance have been studied. It has been shown that the compound possesses a high reaction ability and can be used instead of sodium hydrosulfite in cellulose material dyeing with vat dyes. The suggested method of synthesis can be recommended for further applications.

At present, only two sulfoxylate reducing agents (sodium dithionite = hydrosulfite and sodium hydroxymethylsulfinate = Rongalite) are used in Russia's textile industry.

The main drawback of hydrosulfite consists in the instability of its solutions, especially at high temperatures and high pH-values [1]. This instability leads to the over-use of hydrosulfite in the dyeing process. The known methods of stabilizing hydrosulfite do not permit the complete resolution of this problem. Rongalite is stable in solutions, but it displays its redox properties only at temperatures exceeding 100^0C. But even at high temperatures, the reduction rate of Rongalite is much less than the reduction rate of hydrosulfite [2]. That is why it is impossible to use Rongalite in continuous textile dyeing processes using vat dyes.

For the consumer, the best reducing agents are those which manifest high reducing activity at temperatures below 100^0C and which are resistant to decomposition.

The set of sodium hydroxymethylsulfinate homologs and their amino-derivatives have been synthesized (in laboratory conditions) according to the well-known procedure that includes bisulfite reduction with thermal zinc dust in the presence of aromatic and aliphatic aldehydes. Several physicochemical properties of the synthesized compounds have been studied [3].

However, serious obstacles exist in the way of such a synthesis in industrial conditions. The main difficulty is the shortage of suitable aldehydes (acetic and propyl).

The present work is aimed first at the synthesis of sodium hydroxy-iso-propylensulfate (HiPS) in which the lowest ketone (acetone) is used as an organic stabilizer and secondly at the investigation of the properties of this newly synthesized substance.

Our interest in acetone is connected with the following:

1. Sodium sulfinate, having been stabilized by acetone, is expected to display its highest reaction ability at much lower temperatures in comparison with, for example, hydroxymethylsulfinate and sodium hydroxyethylsulfinate (HES). Such intercorrelation is due to the strongest positive influence of I-induction effect in the sodium hydroxy-iso-propylensulfinate molecule. The influence of this effect results in a C - S -bond energy decrease in HiPS.

2. Acetone is produced in large amounts.

Experiment and discussion

IR-spectroscopic investigations have been done using a Specord M 80 IR-spectrophotometer and with pressed tablets made of KBr. The kinetic data have been obtained using a Specol-211 with a thermostating cell. Redox potentials of the synthesized compounds have been determined with the aid of an electrode pair consisting of platinum and calomel electrodes attached to an OP 211/1 potentiometer.

The synthesis has been performed in a three-necked flask with a stirrer rod, thermometer, and reflux condenser. The liquid components were introduced into the reactive system with the help of a drop funnel. Bisulfite was added to zinc pulp ($Zn:H_2O$ = 1:3, wt %) which had been cooled down to 20^0C and thereafter stirred vigorously for 10 minutes. During the experiment, the temperature did not exceed 35^0C. Then acetone was added, and heating began. The synthesis was carried out for 1.5 hours at a temperature of $90\text{-}93^0C$. Next, the mixture was cooled and filtered.

The mole ratios of the components were as follows: $NaHSO_3:Zn: (CH_3)_2CO = 1 : 1.36 : 1$.

The liquid product was analyzed for HiPS content using the iodometric method [1].

Then water vacuum evaporation at room temperature was carried out at P = 1 mmHg.

The product, having been obtained after water evaporation and air-drying, was analyzed by IR-spectroscopy (Table 1). The data confirmed that the substance obtained (sodium hydroxy- iso-propylsulfinate) belongs to the group of sulfine acid derivatives because two groups of absorption bands were noticed in its IR-spectra. One band characterizes C – S - group absorption while the other characterizes the irregularity of S – O - bonds in the synthesized substance [4].

The reducing activity of the synthesized compound was evaluated by measuring the redox potential (ROP) relative to the calomel electrode.

The ROP-values of sodium hydroxy-iso-propylsulfinate are presented in Table 2 together with the ROP-values for the other sulfoxylate reducing agents. The data demonstrate that HiPS possesses high reducing potential within a temperature interval of 30-90^{0}C.

Table 1.

IR - data for HiPS

Chemical formula of HiPS	Wave numbers (cm^{-1}) at the absorption maxima characteristic of the following groups			
	S–O; S=O	C–C; CH_3	OH–	C–S
$(CH_3)_2OHCSO_2$ Na	530; 580 910; 960 1010; 1090 1050	1050; 1350 1380; 1420 1460; 2800 2900	1200 1294 3600	730 774

The compound obtained can be used as a reducing agent in cellulose-containing material dyeing with vat dyes. Its reducing ability is estimated by the definition of the reducing rate of Indigo Carmine ($IC(CO)_2$).

Table 2.

Redox potentials of several hydroxyalkylsulfinates

Compound	-ROP, mV. Reducing agent content - 6 g/l; pH = 13.5		
	30^0C	60^0C	80^0C
Dithionite	810	985	1060
Rongalite	410	800	900
HES	712	880	922
HiPS	805	960	1003

The formula of the reaction between $IC(CO)_2$ and HiPS in the alkaline medium can be written in a similar manner to those between $IC(CO)_2$ and sodium hydroxymethylsulfinate [5].

$$HO(CH_3)_2SO_2^- \underset{k_{-1}}{\overset{k_1}{\rightleftarrows}} (CH_3)_2CO + HSO_2^-$$

$$IC(CO)_2 + HSO_2^- \underset{k_{-2}}{\overset{k_2}{\rightleftarrows}} IC(COCOH)^- + SO_2$$

The data in Table 3 permit the comparison of the reaction ability of sodium hydroxy-iso-propylsulfinate with those of several other sulfoxylate reducing agents. As is seen in Table 3, the first-order rate constant (having been defined graphically according to the linear anamorphosis of HiPS - $IC(CO)_2$ interaction process kinetic curves) for sodium hydroxy-iso-propylsulfinate is 72 times greater than that for sodium hydroxymethylsulfinate and 2.3 times greater than that for sodium hydroxyethylsulfinate.

Table 3.

The effective rate constants of the reactions between the reducing agents and Indigo Carmine

pH = 8.15; 323 k (50^{0}C)

Compound	Rongalite	HES*	HiPS	Sodium dithionite
K_{ef}, min^{-1} (Reducing agent content - 7.5*10^{-3} mol/l)	$2.50 \cdot 10^{-4}$	$9.03 \cdot 10^{-3}$	$2.06 \cdot 10^{-2}$	$8.0 \cdot 10^{-1}$

* - HES has been synthesized according to the procedures described in [3]

The stability of the alkaline solutions of the discussed compounds to oxidizing decomposition is characterized by the data in Table 4.

Table 4.

The stability of the alkaline solutions of several reducing agents

pH = 13.5; 293 K (20^{0}C)

Compound	Resistance to oxidizing decomposition (%) [1]	
	24 hours	48 hours
Rongalite	100	98
HES	100	94
HiPS	99	96
Sodium dithionite	10	0

The results thus obtained enable us to conclude that the new reducing agent (sodium hydroxy-iso-propylsulinate) possesses the merits of both Rongalite and hydrosulfite. At the same time, HiPS is almost completely devoid of the drawbacks inherent in the two traditional reducing agents. Therefore, HiPS can be recommended as a worthy alternative to hydrosulfite in continuous and discontinuous textile dyeing processes using vat dyes.

References

1. V.V. Budanov, The Chemistry and Technology of Reducing Agents Synthesized on the Base of Sulfoxyl Acid (Moscow: Khimiya, 1984) (in Russian).

2. P.P. Yakimtchuk, A.V. Mishchenko and N.E. Bulusheva, Vat Dye Application (Moscow: Legprombytizdat, 1985) (in Russian).

3. S.V. Makarov, The Stability and Reaction Ability of Sodium Hydroxy- and Aminoalkylsulfinates (Dissertation for the degree of Candidate of chemical science, Ivanovo, 1986)148 (in Russian).

4. A.L. Smith, Applied IR-Spectroscopy: Bases, Technique, Analytical Application (Moscow: Mir, 1982) (Russian Edition).

5. V.V. Budanov and S.V. Makarov, The Chemistry of Sulfur-Containing Reducing Agents (Moscow: Khimiya, 1984) (in Russian).

The Institute of Chemistry of Non-Aqueous Solutions
of the Russian Academy of Sciences, Ivanovo

VINYL ALCOHOL AND VINYL ACETATE COPOLYMERS FOR COTTON YARN SIZING

G.A. Smirnov, V.A. Kurilova, E.A. Indeikin, N.V. Rybin,
M.M. Mogilevich, E.V. Molchanova

The sizing properties of vinyl alcohol and vinylacetate copolymers possessing different contents of vinylacetate links have been investigated. These copolymers have been obtained by acid hydrolysis of polyvinylacetate in aqueous dispersion. It has been established that cotton yarn sized with the mentioned new reagents is characterized by improved physico-mechanical properties in comparison with starch sized yarn. The created copolymers are shown to be applicable for "cold " yarn sizing.

In an earlier publication [1], we have shown that the water-soluble copolymers of vinyl alcohol (VAl) and vinylacetate (VAc) (produced by polyvinylacetate (PVAc) acidic hydrolysis in water dispersion) ensure better fastness properties of linen yarn in comparison with starch.

In the present work, the technological properties of VAl and VAc copolymer aqueous solutions and their application in cotton yarn sizing were investigated. These copolymers, containing different amounts of vinylacetate links, have also been obtained by PVAc acetic hydrolysis.

The process of VAl and VAc water-soluble copolymer production was catalyzed by ortophosphatic acid in the presence of sodium alkylbenzensulphonate (sulphonol) in conformity with the procedure described in [1]. The following reagents were used: mean-viscose unplastified coarse grained PVAc dispersion D51S (GOST 18992-80), oriophosphatic acid (GOST 10678-7), and sulphonol (TY 6-01-1043-86). The hydrolysates were purified of ortophosphatic and acetic acids by dialysis using cellophane as a membrane. Yarn sizing was performed on a laboratory sizing machine that permitted contact drying to be done. Cotton yarn with a linear density of 29.0 tex was applied as a substrate.

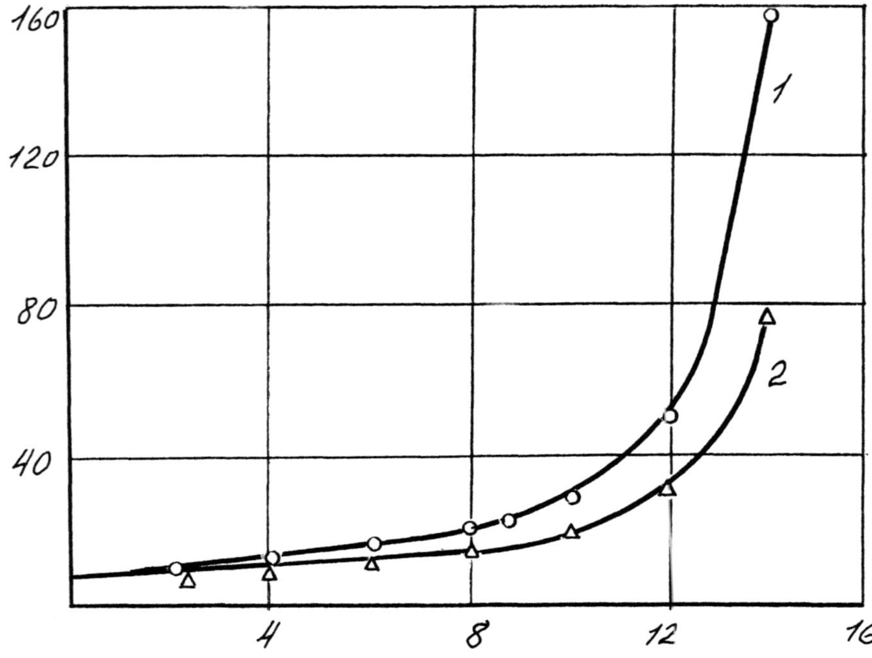

Fig. 1. The dependence of VAl and VAc copolymer aqueous solution viscosity on the copolymer mass portion at 24^{0}C. VAc link content in the copolymers: 1 - 12.3%; 2 - 32.8%.. X - copolymer mass portion; Y - conditional viscosity, s*
* - measured with a ÂÇ-264-4 viscosimeter

According to the data in Fig.1, the inflection of the curve, describing the dependence of VAl and VAc copolymer aqueous solution viscosity upon the copolymer concentrations, is observed when the mass portion of copolymers is equal to 10-12%. This regularity is connected with the physical net formation (structurizing of the solution). It must be noted that within the concentration interval of 2-10% at room temperature, the aqueous solutions of the copolymers under investigation are characterized by low viscosity values (0.05-0.15 Pa·s). The change in copolymer remnant VAc link content from 12% to 32% influences the viscosity indices significantly.

The results of the research into the properties of cotton yarn sized with VAl and VAc copolymers, potato starch, and polyvinyl alcohol (PVAl) are presented in Table 1. The data reveal that VAl and VAc copolymers posses excellent sizing properties.

The application of these reagents provides an increase in yarn tensile strength of 27.5-33.0% (depending on the paste value), a rise in abrasion resistance of 20-34%, and an improvement in multiple elongation resistance of 118-159%. The indices achieved by the use of polyvinyl alcohol for yarn sizing are comparable with those which have been just represented (with the exception of yarn abrasion resistance characteristics). A starch sized yarn, if compared with yarn sized with VAl, VAc, and PVAl, is worse in physico-mechanical properties even at high paste values.

Table 1.

Yarn characteristics	Soft yarn	Polymer, used for sizing: VAl and VAc copolymer (15.2% of VAc links) with the mass portion, %			Polyvinyl alcohol with the mass portion, %			Starch with the mass portion, %
		2	3	4	2	3	4	5
True paste, %	-	2.6	3.7	4.3	2.6	3.7	4.2	4.9
Linear density, tex	28.6	30.9	29.8	30.0	29.8	29.0	29.8	30.6
Increase in tensile strength in comparison with soft yarn, %	-	27.5	30.0	33.0	29.0	33.0	44.2	23.4
Variation coefficient of tensile strength, %	14.3	10.5	12.3	12.5	10.2	10.2	9.8	12.1
Increase in abrasion resistance in comparison with soft yarn	-	-	20.2	34.2	13.3	169.0	176.8	24.3
Increase in multiple elongation resistance in comparison with soft yarn, %	-	118.4	159.1	153.1	104.6	156.5	156.2	107.5
Hardness coefficient	4.8	10.6	12.2	13.9	10.5	12.4	16.8	8.3

Fig. 2 illustrates the dependence of the relative increase in yarn tensile strength after sizing upon the VAc link content in the copolymers. The relative increase in yarn tensile strength has been calculated according to the equation given in [1]. As seen in Fig. 2, the yarn sized with VAl and VAc copolymers (in which the concentration of VAc links lies between 15 and 35%) is characterized by a larger relative increase in tensile strength if compared with the starch sized yarn. The copolymers containing 15-23% of VAc links are preferable because their application enables an increase in yarn tensile strength of up to 27-19%. These high tensile strength values correspond to the low values of true paste

(2.6-2.7%). The copolymer mass portion in the sizing compositions was equal to 2.0-2.6%.

Temperature plays a great role in yarn sizing. For starch and polyvinyl alcohol sizing compositions the optimal temperature of the yarn sizing processes is about 85-90^{0}C. For VAl and VAc copolymers, it has been established that their aqueous solution viscosity increases very slightly in the course of temperature lowering. Thus, the diminution of temperature from 90 to 24^{0}C causes a rise in viscosity from 0.043 Pa·s to 0.55 Pa·s (for solutions with a copolymer mass portion of 4%). This fact allows us to propose that "cold " sizing is possible if the developed sizing copolymers will be used. To check this hypothesis,, yarn sizing processes have been performed at temperatures of 86 and 25^{0}C using a VAl and VAc copolymer pattern containing 22.3% of VAc links. As follows from the data in Table 2, the VAl and VAc copolymer ensures an equal increase of yarn tensile strength at both temperatures (86 and 25^{0}C). Yarn sizing with polyvinyl alcohol gives worse results at 25^{0}C than at 86^{0}C. For example, the increase in yarn tensile strength at 86^{0}C is 1.7 times greater in comparison with the yarn tensile strength increase at 25^{0}C.

Table 2.

Sizing parameters and yarn indices	VAl and VAc copolymer (22.3% of VAc links)		Polyvinyl alcohol	
Sizing composition temperature, ^{0}C	86	25	86	25
Copolymer mass portion in the sizing composition, %	2.7	2.6	2.5	2.6
True paste, %	3.0	2.5	2.7	2.6
Yarn tensile strength increase in comparison with soft yarn, %	19.2	19.9	31.0	18.0

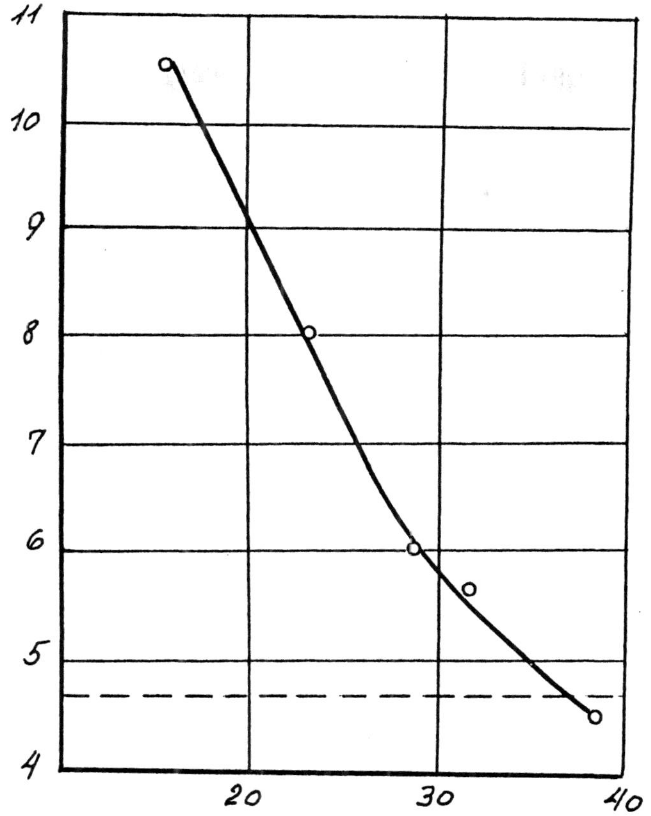

Fig. 2. The dependence of the relative increase in sized cotton yarn tensile strength on VAc link content in VAl and VAc copolymers. Dotted line = the relative increase level in tensile strength of the cotton yarn sized with potato starch. X - VAc link content in the copolymers, %; Y - the relative increase in tensile strength.

Thus, the consideration of water-soluble VAl and VAc copolymer properties (these, as was already mentioned, have been obtained by PVAl acidic hydrolysis in aqueous dispersion and purified by dialysis) has shown that their application in yarn sizing provides better yarn tensile strength, multiple elongation resistance, and abrasion resistance than starch. It has been demonstrated that the sizing properties of VAl and VAc copolymer aqueous solutions are not dependent on temperature, and hence, "cold " yarn sizing is quite feasible.

References

1. G.A Smirnov et al., Textile Chemistry, (1) (1994) 33 (in Russian).

The State Technical Univercity, Yaroslavl;
The Scientific Research Institute of Cotton Industry, Ivanovo

Part III

The Theoretical Bases of Chemical Technology of Textile Material

DESCRIPTION OF KINETIC REGULARITIES OF DYEING PROCESS BY THE USE OF SORPTION CURVE MOMENTS

F.Yu.Telegin, A.V.Shormanov, B.N.Mel'nikov

The publications in the literature have been analysed where the kinetic curve moments have been used as the integral characteristic of sorption process. This article shows the development of moment-using methodology to describe the processes of textile dyeing. The regularities of application of sorption curve moments to the analysis of the mechanisms of complex processes, often indescribable by other analytical methods, are discussed. The considerations of the kinetic regularities of fibrous material package dyeing serve an illustration of the approach proposed. Examples are given of the practical application of the developed method to create an effective technology of textile material dyeing.

As a rule, the sorption rate distribution moments [1-5] (called for short "kinetic curve moments") are used as independent integral characteristics of sorption process provided that the transfer coefficients are constant. These moments are used to analyse its regularities and to calculate the kinetic constants according to the experimental data. But consideration of transport phenomena requires the solution of more general problems connected with the definition of sorption kinetic curves. Though a great number of analytical methods of solving such problems have been developed, at present the effective algorithms have not been found using moments. At the same time, well-known methods of restoration of distribution by the use of moments [1, 6-8] can be the theoretical premises for their application as integral transformation while solving the kinetic tasks.

The sorption rate distribution moments are mostly defined by the solution of sorption problem in Laplace space [2, 3]. The limited possibilities to apply this procedure to the calculation of high-order moments hamper its wide spread use. Only the equations for the first three moments are usually given. The general expression for the moments of random order, convenient for calculations, are obtained only in some special cases [2], such as the diffusion in bi-porous sorbent with a right-angled sorption isotherm and with the prescribed kinetic function of micro-pores.

The more effective method of moment definition is based on the transformation of differential equations in partial derivatives, which describe the alterations of sorbate concentration in time and space, into the system of ordinary differential equations [1, 5-6], reflecting space distribution of sorption rate moments. But the potentialities of this method have not been used utterly to the estimation of high-order sorption curve moments.

The description of sorption by homogeneous small-porous sorbent in the shape of infinite plate, round cylinder, or sphere leads to the necessity of solving the equation

$$\frac{\partial a}{\partial t} = \frac{D}{x^{\nu}}\frac{\partial}{\partial x}\left(x^{\nu}\frac{\partial a}{\partial x}\right), \quad t>0, \quad a \equiv a(x,t) \qquad (1)$$

with conditions

$$\partial a(0,t)/\partial x = 0, \quad a(x,0) = a_0 = Const \;, \qquad (2)$$

where a = local sorbate concentration; x = space co-ordinate; t = time; D = sorbate diffusion coefficient; ν = sorbent shape coefficient (ν = 0 for a plate, ν = 1 for a cylinder; ν = 3 for a sphere).

The integral correlation between the variables a(x,t) and $m_n(x)$

$$m_n(x) = \frac{\left(D/R^2\right)^n}{n!}\int_0^{\infty} t^n \frac{\partial a(x,t)}{\partial t}dt \qquad (3)$$

used with method [1] permits the conversion of equations (1)-(2) into the system

$$-m_{n-1} = \frac{1}{z^{\nu}}\frac{d}{dz}\left(z^{\nu}\frac{dm_n}{dz}\right), \quad m_n \equiv m_n(z), \quad n=1,2,3...$$

$$dm(0)/dz = 0, \quad m_0 = a(z,\infty) - a(z,0) = Const, \qquad (4)$$

$$z = x/R$$

where R is a half of the plate thickness or the radius of the cylinder or sphere.

Formula (3) coincides with the definition of sorption rate distribution moment within a multiplier [1, 2]. The use of the factorial multiplier in the traditional statement of moments does not affect the meaning of integral transformation. That is why the adopted term "moment" will be used in this article instead of the special notion of "inertia" proposed in [6].

Taking into account the conditions of symmetry, the general solution of the system (4) can be written as [9]

$$m_n(z) = \sum_{k=0}^{n}(-1)^k B_{n-k} \frac{\Gamma(k+1/2)(z/2)^{2k}}{k!\Gamma(k+\nu/2+1/2)}, \qquad (5)$$

where $B_n = m_n(0)$ is the integration constant.

To make the calculations more convenient, it is possible to organise the equation (5) in the following form (taking into account the peculiarities of Γ-function):

$$m_n(z) = B_n + \sum_{k=0}^{n-1}(-1)^k B_{n-k} \frac{z^{2k}}{2^k k!(\nu+1)(\nu+3)...(\nu+2k-1)}. \qquad (6)$$

The mean value concentration of sorbate is the most important parameter in solving sorption problems:

$$A(t) = \frac{\nu+1}{R^{\nu+1}}\int_0^R a(x,t)x^\nu dx = (\nu+1)\int_0^1 a(z,t)z^\nu dz. \qquad (7)$$

The corresponding general rate sorption distribution moments are given by the formula

$$M_n = \frac{\left(D/R^2\right)^n}{n!}\int_0^\infty t^n \frac{\partial A(t)}{\partial t}dt = (\nu+1)\int_0^1 m_n(z)z^\nu dz. \qquad (8)$$

If equation (6) will be taken into consideration, the general rate sorption moments gain the following form:

$$M_n = B_n + \sum_{k=1}^{n}(-1)^k B_{n-k} \frac{1}{2^k k!(\nu+3)(\nu+5)...(\nu+2k+1)}, \qquad (9)$$

where $M_0 = m_0 = B_0$.

The integration constants B_n, necessary for the final definition of moments, must correspond to the boundary conditions for the system of differential equations (4) and for the initial differential equation (1).

In order to illustrate and test the proposed method of calculation of high-order moments, it is expedient to consider the description of diffusion kinetics sorption from the limited volume taking into account external diffusion in the course of sorbate transfer to the sorbent surface. In this case, the boundary conditions can be given by the equations

$$\frac{(\nu+1)D}{R}\frac{\partial a(R,t)}{\partial x}=\alpha\left[C(t)-\frac{a(R,t)}{\Gamma}\right] \qquad (10)$$

$$C(t)+\frac{A(t)}{\lambda_V}=C_0,\quad \lambda_V=\frac{V_s}{V_f} \qquad (11)$$

where C_0 and C(t) = the initial and current concentrations of sorbate in the external volume; a(R,t) = the sorbate concentration on the sorbent surface; V_f and V_s = the fibre and solution volume; α = the mass-transfer coefficient. This coefficient characterises the density of the substance flow through the phase-separating boundary and relates to the sorbent volume of one measurement unit provided that the sorbate concentration gradient in the external medium is also equal to unity; G = the partition coefficient.

If λ_V has a limited value, then it is suitable to express the mean value concentration A(t) as a function of sorbate exhaustion degree E(t) , bearing in mind the balance ratio (11)

$$A(t)=C_0\lambda_V E(t),\quad E(t)=1-\frac{C(t)}{C_0}. \qquad (12)$$

The kinetic problem presented by equations (1), (2), (7), (10)-(12) has a strict analytical solution as given in monograph [10].

The moment boundary conditions are as follows

$$-\frac{dm_n(1)}{dz}+Bi\left[\Gamma M_{C,n}-m_n(1)\right]=0$$

$$z=\frac{x}{R},\quad Bi=\frac{\alpha R}{(\nu+1)\Gamma D} \qquad (13)$$

$$M_{C,n}+\frac{M_n}{\lambda_V}=0$$

where $M_{C,n} = \dfrac{\left(D/R^2\right)^n}{n!} \int_0^\infty t^n \dfrac{dC(t)}{dt} dt$,

Bi = Bio's number.

The developed general solution (6) and (9) permits the integration constant for boundary and balance conditions (13) to be defined as

$$B_n = -\frac{1}{1+\dfrac{\Gamma}{\lambda_V}} \left[\sum_{k=1}^{n} (-1)^k B_{n-k} \left(g_k + \frac{1}{Bi} g'_k + \frac{\lambda_V}{\Gamma} G_k \right) \right]$$

$$g_k = \frac{1}{2^k k!(\nu+1)(\nu+3)\dots(\nu+2k-1)}$$

$$g'_k = \frac{1}{2^{k-1}(k-1)!(\nu+1)(\nu+3)\dots(\nu+2k-1)} \qquad (14)$$

$$G_k = \frac{1}{2^k k!(\nu+3)(\nu+5)\dots(\nu+2k+1)}$$

$$B_0 = M_0 = A(\infty) - A(0) = 1$$

The different methods of determination restoration in accordance with the integral characteristics [1, 6-8] can be used to solve sorption problems applying the discussed high-order moments. As the analysis has shown [9], those methods which use the interconnection between distribution moments and the representation of distribution function in the form of Laplace integral transformation appeared to be the most effective. The mentioned interconnection if it is used to describe the sorption rate distribution function $dA(\tau)/d\tau$, $\tau = Dt/R^2$ can be written as

$$\int_0^\infty \exp(-s\tau) \frac{\partial A(\tau)}{\partial \tau} d\tau = \int_0^\infty \frac{\partial A(\tau)}{\partial \tau} \sum_{k=0}^{\infty} (-1)^k \frac{(s\tau)^k}{k!} d\tau = \sum_{k=0}^{\infty} (-1)^k M_k s^k \quad .$$

(15)

To realise the possibilities of high-order moment calculations while approximating the sorption kinetic curve, let us apply the well-known method of Laplace transformation inversion accompanied by the expansion of distribution function representation (15) into the sum of residues in the set of its poles [11, 12]. Such an approach will give the opportunity to solve the kinetic problem in the form of terminal exponential sequence, approximating the exact solution in the best way

$$\sum_{k=0}^{\infty}(-1)^k M_k s^k = \frac{a_0 s^n + a_1 s^{n-1} + a_2 s^{n-2} + \ldots + a_{n-1}s + a_n}{s^n + b_1 s^{n-1} + b_2 s^{n-2} + \ldots + b_{n-1}s + b_n} = \frac{P(s)}{Q(s)} . \quad (16)$$

Thus the coefficients b_1 , b_2 , ...,b_n present the solution of the linear algebraic equation system [9, 13]:

$$\begin{vmatrix} -M_1 & M_2 & \cdots & (-1)^n M_n \\ M_2 & -M_3 & \cdots & \\ \cdots & \cdots & \cdots & \cdots \\ (-1)^n M_n & (-1)^{n+1} M_{n+1} & \cdots & (-1)^{2n-1} M_{2n-1} \end{vmatrix} \begin{vmatrix} b_1 \\ b_2 \\ \cdots \\ b_n \end{vmatrix} = - \begin{vmatrix} M_0 \\ M_1 \\ \cdots \\ M_{n-1} \end{vmatrix} , \quad (17)$$

and the coefficients a_1 , a_2 ,...,a_n are defined according to the relation

$$a_m = \sum_{k=0}^{n-m} (-1)^k M_k b_{k+m} , \quad (18)$$

using the known values of b_1 , b_2 ,...,b_n for m = 1, 2, ... , n-1.

In addition, it is necessary to remember that the coefficients a_n and b_n must satisfy the properties of the kinetic curve $A(\tau)$ both at the initial moment of time $\tau = 0$ and the equilibrium state $\tau \to \infty$, i.e.

$$a_0 = 0, \quad a_n / b_n = A(\infty) .$$

The application of the inverse Laplace transformation formula [11] to the expression (16) permits the kinetic curve sought for to be calculated

$$\frac{A(\tau)}{A(\infty)} = 1 + \sum_{k=1}^{\infty} \frac{P(s_k)}{s_k Q'(s_k)} \exp(s_k \tau), \quad \tau = \frac{Dt}{R^2} , \quad (19)$$

where s_k = the roots of the characteristic equation

$$s^n + b_1 s^{n-1} + b_2 s^{n-2} + \ldots + b_{n-1}s + b_n = 0 , \quad (20)$$

$P(s_k)$ = the value of the rational function (16) numerator at the point where $s=s_k$; $Q'(s_k)$ = the derivative of the rational function (16) denominator at the same point.

Calculation of the roots of the equation (20) may be reduced to the definition of Hessenberg matrix eigen-values using well-known numerical algorithms of linear algebra [14].

The results [15] demonstrated that the application of moments enables us to describe sorption kinetics from the unlimited external volume within an error of 2.5% for each part of the curve. The optimal variant of the approximating formula contains four exponents and, as a result, demands the calculation of the first seven moments of the kinetic curve. The determination of sorption kinetics from the limited volume with an error within 5% can be achieved for the values of relative adsorption $A(t)/A(\infty) > 0.5$. In this case, the accuracy of the procedure decreases as the equilibrium exhaustion degree $E(\infty)$ of sorbate from the external medium increases. The optimal variant of the approximating formula does not exceed five exponents.

For practical purposes, the calculations can be made using micro-calculators if the simplified formulas are applied. Thus, the definition of the first three sorption rate distribution moments permits us to describe the kinetic curve by means of the equation

$$\frac{A(\tau)}{A(\infty)} = 1 + \sum_{k=1}^{2} \frac{a_1 s_k + a_2}{s_k (2s_k + b_1)} \exp(s_k \tau), \quad \tau = \frac{Dt}{R^2}$$

$$s_{1,2} = \frac{b_1}{2} \pm \sqrt{\frac{b_1^2}{4} - b_2} \qquad (21)$$

$$b_1 = \frac{M_1 M_2 - M_3}{M_2^2 - M_1 M_3}, \quad b_2 = \frac{M_1^2 - M_2}{M_2^2 - M_1 M_3}$$

$$a_1 = b_1 + M_1 b_2, \quad a_2 = b_2$$

The application of the single first moment makes the equation much simpler:

$$\frac{A(\tau)}{A(\infty)} = 1 - \exp\left(\frac{Dt}{R^2}\frac{M_1}{m_0}\right), \quad m_0 = A(\infty) \; . \qquad (22)$$

In the conformity with the discussed mathematical description of sorption kinetics from the limited solution volume by means of equations (1), (2), (7), (10), (11) the first moment is defined by the formula

$$\frac{M_1}{m_0}=\frac{1}{1+\frac{\lambda_V}{\Gamma}}\left(\frac{1}{(\nu+1)(\nu+3)}+\frac{1}{(\nu+1)Bi}\right). \qquad (23)$$

The result obtained illustrates visually that the value of the first moment defines the dimensionless time for the achievement of 63% of dye yield on the fibre. The same formula together with (8) shows that the variable, inverse to the first sorption rate distribution curve moment, describes the rate of the process. The revealed physical meaning of the moment M_1/m_0 opens the possibility to use this parameter as an integral characteristic for the analysis of the kinetic regularities of dyeing processes. A similar analysis positioned on the intercorrelation between the equilibrium and the kinetic phenomena in the boundary layer and inside the fibre has already been done elsewhere [10, 16, 17] in conformity with continuous and periodic processes.

Thus the presented results confirm the possibility of the application of the method of moments for the solution of diffusion kinetic problems and enables us to pass on to the consideration of more complex processes of practical interest.

One of these processes is the process characterised by the simultaneous phenomena of diffusion and first-order chemical reaction. Earlier in article [18], while solving this problem, the formulas for the first two moments were derived and these are insufficient to restore the kinetic curve.

In the course of describing the diffusional sorption process, accompanied by an irreversible chemical reaction, taking place in the small-porous sorbent in the shape of an infinite plate, a round cylinder, or a sphere, the necessity arises to solve the equation [19]

$$\frac{\partial a}{\partial t}=\frac{D}{x^{\nu}}\frac{\partial}{\partial x}\left(x^{\nu}\frac{\partial a}{\partial x}\right)-\frac{\partial f}{\partial t}$$

$$\frac{\partial f}{\partial t}=k_f a, \quad t>0, \qquad (24)$$

$$a \equiv a(x,t), \quad f \equiv f(x,t)$$

with the conditions

$$\partial a(0,t)/\partial x=0, \quad a(x,0)=a_0=Const \qquad (25)$$

where a = the diffusing sorbate local concentration; f = the local concentration of sorbate having come into the irreversible interaction with sorbent; k_f = the rate constant of the irreversible first order chemical reaction.

The condition given by (24) corresponds to the sorption from an external medium with limited volume when (due to the irreversibility of the chemical reaction) the sorbate, initially introduced into the external volume, is expended entirely and the uniform (zero) terminal distribution of the diffusing substance is achieved.

In the procedure of calculating moments

$$m_{a,n}(z) = \frac{1}{n!}\int_0^\infty \tau^n \frac{\partial a(z,\tau)}{\partial \tau} d\tau$$
$$m_{f,n}(z) = \frac{1}{n!}\int_0^\infty \tau^n \frac{\partial f(z,\tau)}{\partial \tau} d\tau \tag{26}$$

after the replacement of variables $z = x/R$, $\tau = Dt/R^2$ and using method [1] the equations (24), (25) are transformed into the system

$$-m_{a,n-1} = \frac{1}{z^\nu}\frac{d}{dz}\left(z^\nu \frac{dm_{a,n}}{dz}\right) - b^2 m_{a,n}$$
$$-m_{f,n-1} = b^2 m_{a,n}$$
$$m_{a,n} \equiv m_{a,n}(z), \quad m_{f,n} \equiv m_{f,n}(z), \quad n = 1,2,3... \tag{27}$$
$$\frac{dm_{a,n}(0)}{dz} = 0, \quad m_{a,0} = a(z,\infty) - a(z,0) = -a_0 = Const$$
$$b^2 = \frac{k_f R^2}{D}$$

The boundary condition for the sorbate surface will be discussed later when solving the concrete sorption task.

The general solution of the problem (27) may be represented as follows [20]:

$$
\begin{aligned}
m_{a,n}(z) &= \sum_{k=0}^{n-1} \frac{(-1)^k B_{n-k}}{\Gamma\left(\tfrac{1}{2}+\tfrac{\nu}{2}+k\right)} \frac{(z/2)^{2k}}{k!} \sum_{i=0}^{\infty} \frac{(bz/2)^{2i}}{\Gamma\left(\tfrac{1}{2}+\tfrac{\nu}{2}+k+i\right) i!} + \\
&+ (-1)^n B_0 \Gamma\left(\tfrac{1}{2}+\tfrac{\nu}{2}\right) \frac{(z/2)^{2n}}{n!} \sum_{i=0}^{\infty} \frac{n}{n+i} \frac{(bz/2)^{2i}}{\Gamma\left(\tfrac{1}{2}+\tfrac{\nu}{2}+n+i\right) i!} = \\
&= \sum_{k=0}^{n-1} (-1)^k B_{n-k} \frac{\Gamma\left(\tfrac{1}{2}+\tfrac{\nu}{2}\right)}{\Gamma\left(\tfrac{1}{2}+\tfrac{\nu}{2}+k\right)} \frac{(z/2)^{2k}}{k!} \; {}_0F_1\left(\begin{matrix} \tfrac{1}{2}+\tfrac{\nu}{2}+k; \\ b^2 z^2/4 \end{matrix}\right) + \\
&+ (-1)^n B_0 \frac{\Gamma\left(\tfrac{1}{2}+\tfrac{\nu}{2}\right)}{\Gamma\left(\tfrac{1}{2}+\tfrac{\nu}{2}+n\right)} \frac{(z/2)^{2n}}{n!} \; {}_1F_2\left(\begin{matrix} n; \; b^2 z^2/4 \\ 1+n; \tfrac{1}{2}+\tfrac{\nu}{2}+n \end{matrix}\right)
\end{aligned} \tag{28}
$$

$$
m_{f,n-1}(z) = -b^2 m_{a,n}(z) \tag{29}
$$

The mean volume concentrations of the diffusing A(t) and chemically bonded F(t) sorbate are defined by the relations

$$
A(\tau) = (\nu+1)\int_0^1 a(z,\tau) z^{\nu+1} dz, \quad F(\tau) = (\nu+1)\int_0^1 f(z,\tau) z^{\nu+1} dz \; . \tag{30}
$$

The corresponding sorption rate distribution moments

$$
M_{a,n} = \frac{1}{n!}\int_0^{\infty} \tau^n \frac{dA(\tau)}{d\tau} d\tau, \quad M_{f,n} = \frac{1}{n!}\int_0^{\infty} \tau^n \frac{dF(\tau)}{d\tau} d\tau \tag{31}
$$

can be determined according to the following formulas provided that expressions (28) and (29) are taken into consideration

$$
\begin{aligned}
M_{a,n} &= \sum_{k=0}^{n-1} (-1)^k B_{n-k} \frac{\Gamma\left(\tfrac{3}{2}+\tfrac{\nu}{2}\right)}{\Gamma\left(\tfrac{3}{2}+\tfrac{\nu}{2}+k\right)} \frac{1}{4k!} \; {}_0F_1\left(\begin{matrix} \tfrac{3}{2}+\tfrac{\nu}{2}+k; \\ b^2/4 \end{matrix}\right) + \\
&+ (-1)^n B_0 \frac{\Gamma\left(\tfrac{3}{2}+\tfrac{\nu}{2}\right)}{\Gamma\left(\tfrac{3}{2}+\tfrac{\nu}{2}+n\right)} \frac{1}{4n!} \; {}_1F_2\left(\begin{matrix} n; \; b^2/4 \\ 1+n; \tfrac{3}{2}+\tfrac{\nu}{2}+n \end{matrix}\right)
\end{aligned} \tag{32}
$$

where

$$
M_0 = m_0 = B_0 \; .
$$

At the final stage of the definition of moments, the integration constant must be calculated in accordance with the boundary conditions for both the moment equation system (27) and the initial differential equation (24). As it was mentioned, the boundary condition (25) demands that the external volume (from which the sorbate transfer takes place) be limited. Additionally, it is reasonable to take into consideration the external diffusional resistance in the course of sorbate flow towards the sorbent surface and the procession of a first-order irreversible reaction between the sorbate and the solvent in the external volume resulting in the loss of sorbate reactivity [18].

Such conditions are realised, for example, in the sorption process of reactive dyes which chemically interact with protein and cellulosic substrates and with water [21, 22]. The analysed process of simultaneous diffusion and chemical reaction, taking place inside the sorbent, is more typical for wool dyeing [23].

Thus the missing boundary conditions for the solution of the problem (24)-(25) can be represented in the following way

$$\begin{aligned}
&\frac{(\nu+1)D}{R}\frac{\partial a(R,t)}{\partial x} = \alpha\left[C(t) - \frac{a(R,t)}{\Gamma}\right] \\
&C(t) + C_w(t) + \frac{A(t)+F(t)}{\lambda_V} = C_0 \\
&\lambda_V = \frac{V_s}{V_f}, \quad C_0 \equiv C(0) \\
&\frac{dC_w(t)}{dt} = k_w C(t), \quad C_w(0) = 0
\end{aligned} \tag{33}$$

Having been transformed into moments, the system assumes the form

$$-\frac{dm_{a,n}(1)}{dz}+Bi\left[\Gamma M_{C,n}-m_{a,n}(1)\right]=0$$

$$Bi=\frac{\alpha R}{(\nu+1)\Gamma D}$$

$$b_w^2 M_{C,n}=\frac{M_{a,n-1}+M_{f,n-1}}{\lambda}+\Gamma M_{C,n-1} \qquad (34)$$

$$\lambda=\frac{V_f}{\Gamma V_s}, \quad b_w^2=\frac{k_w R^2}{D}$$

$$M_{C,n}=\frac{1}{n!}\int_0^\infty \tau^n \frac{dA(\tau)}{d\tau}d\tau$$

where $M_{c,0}$ = - C_0 since C (∝)=0 due to the last equation of the system (33); C and C_0 = the current and initial concentrations of sorbate capable of interacting with a sorbent of volume V_f and being located in the external medium with the volume V_s ; C_w = the current concentration of sorbate, having interacted with the solvent and being unable to react further with the sorbent.

The correlations (28), (29) and (31), (32) permit us to define the general recursion formula for the integration constants and to express in brief the moments:

$$m_{a,n}=\sum_{k=0}^{n}B_{n-k}g_k, \quad m'_{a,n}=\sum_{k=0}^{n}B_{n-k}g',$$

$$M_{a,n}=\sum_{k=0}^{n}B_{n-k}G_k, \quad M_{f,n-1}=-b^2 M_{a,n}$$

$$B_n=\frac{\sum_{k=0}^{n-1}B_{n-k}\left[g_k-b_w^2 g_{k+1}+(g'_k-b_w^2 g'_{k+1})/Bi+(G_k-b_w^2 G_{k+1})/\lambda\right]}{b_w^2(g_0+g'_0/Bi)+b^2 G_0/\lambda}$$

(35)

in which the following designations are introduced

$$g_k = (-1)^k \frac{\Gamma(1/2+\nu/2)}{\Gamma(1/2+\nu/2+k)} \frac{(1/2)^{2k}}{k!} \, {}_0F_1\left(\begin{matrix} 1/2+\nu/2+k; \\ b^2/4 \end{matrix}\right), \quad k<n$$

$$g_n = (-1)^n \frac{\Gamma(1/2+\nu/2)}{\Gamma(1/2+\nu/2+n)} \frac{(1/2)^{2n}}{n!} \, {}_1F_2\left(\begin{matrix} n; \; b^2/4 \\ 1+n; 1/2+\nu/2+n \end{matrix}\right),$$

$$g'_k = (-1)^k \frac{\Gamma(1/2+\nu/2)}{\Gamma(1/2+\nu/2+k)} \frac{(1/2)^{2k-1}}{(k-1)!} \, {}_1F_2\left(\begin{matrix} 1+k; \; b^2/4 \\ k; 1/2+\nu/2+k \end{matrix}\right), \quad k<n$$

$$g'_n = (-1)^n \frac{\Gamma(1/2+\nu/2)}{\Gamma(1/2+\nu/2+n)} \frac{(1/2)^{2n-1}}{(n-1)!} \, {}_0F_1\left(\begin{matrix} 1/2+\nu/2+n; \\ b^2/4 \end{matrix}\right),$$

$$G_k = (-1)^k \frac{\Gamma(3/2+\nu/2)}{\Gamma(3/2+\nu/2+k)} \frac{1}{4k!} \, {}_0F_1\left(\begin{matrix} 3/2+\nu/2+k; \\ b^2/4 \end{matrix}\right), \quad k<n$$

$$G_n = (-1)^n \frac{\Gamma(3/2+\nu/2)}{\Gamma(3/2+\nu/2+n)} \frac{1}{4n!} \, {}_1F_2\left(\begin{matrix} n; \; b^2/4 \\ 1+n; 3/2+\nu/2+n \end{matrix}\right)$$

Thus the last created formulas give the possibility to define the arbitrary number of the rate distribution moments of the diffusional sorption process, accompanied by an irreversible chemical reaction. This permits us to restore the sorption kinetic curves by the use of the methods discussed in this article.

In the special case where the sorbent has the shape of an infinite round cylinder (i.e. when dyeing fibres), the mean time for the achievement of a 63% degree of general sorption completion $M_{s,1} = M_{a,1} + M_{f,1}$ and covalent fixation $M_{f,1}$ of dye can be found according to the formulas

$$M_{s,1} = \frac{C_0 \lambda_V}{b} \frac{Q_1 \lambda}{(1+Q_0\lambda)^2}; \quad M_{f,1} = \frac{C_0 \lambda_V}{b} \left[\frac{Q_1 \lambda}{(1+Q_0\lambda)^2} + \frac{1}{1+Q_0\lambda} \right]$$

$$Q_0 = \frac{b_w}{2} \left(\frac{I_0(b)}{b I_1(b)} + \frac{1}{Bi} \right)$$

$$Q_1 = \frac{b I_0(b)}{I_1(b)} + \frac{b_w^2}{4} \left(\frac{I_0^2(b)}{b I_1^2(b)} - 1 \right) + \frac{b_w^2}{4Bi}$$

(36)

I_0 , I_1 = the modified Bessel's functions of the first kind for the zero and the first numbers.

While describing the kinetics of chemosorption, complicated by dye hydrolysis in a solution, the value of the zero moment is of great interest

$$M_{f,0} = \frac{C_0 \lambda_V}{1 + \frac{Q_0 \lambda_V}{\Gamma}} . \qquad (37)$$

The physical meaning of this parameter corresponds to the finite mean dye concentration in the volume of sorbent. In the literature [22], there exists the notion "efficiency of dyeing", which is similar to (37) in its meaning but different in its mathematical formulation. Earlier, while considering the equilibrium processes, the zero sorption moment was determined by the formula (4). In the case where the first-order chemical reaction of covalent fixation of the dye with the fibre takes place in a bath of limited volume, theoretically the whole dye in the solution must interact with the polymer functional groups [23]. But if the reaction of hydrolysis also occurs, this theoretical dye concentration will reduce according to the formula (37). The peculiarities of this process of simultaneous diffusion and chemical reaction of dyes with sorbents are analysed in detail in the following publications [10, 16, 17, 21-25].

As is known, the serious problem of ensuring the uniform dyeing of textile packages arises while dyeing textiles in a circulation type apparatus. Numerous attempts to solve this problem reveal the necessity for serious theoretical investigations of the regularities of mass-transfer processes taking place in the package. But even today, the complexity of the description of the whole totality of the phenomena, proceeding during package dyeing, presents an obstacle along the path of developing a full mathematical model. In this connection, the application of the method of moments is useful to the analysis of the package dyeing process.

In order to simulate the process discussed , the developed equation system must be supplemented with the equation of convection transfer in the package

$$\frac{\partial C}{\partial t}+\frac{1-\varepsilon}{\varepsilon}\frac{\partial A}{\partial t}=-\frac{1}{X_0+X}\frac{\partial}{\partial X}[(X_0+X)UC]$$

$$U=\frac{U_0X_0}{X_0+X} \qquad (38)$$

as well as the following initial

$$C(X,0)=C_0, \quad A(X,0)=0$$

and boundary conditions

$$C(0,t)=C_s$$

$$C(L,t)=C_s$$

in the course of liquid circulation forwards from the centre of the package to its periphery and in opposite direction.

The balance correlation is written as

$$\frac{2}{L(L+2X_0)}\left(\frac{1-\varepsilon}{\varepsilon}\int_0^L AXdX+\int_0^L CXdX\right)+\frac{V_s}{V_p}C_s=\left(1+\frac{V_s}{V_p}\right)C_0 \,. \qquad (39)$$

In this equation X_0 = the internal radius of the package; X = the current coordinate, beginning at X_0 ; L = the thickness of the winding layer of the package; U_0 = the liquid flow speed in the layer at X_o; U = the liquid flow speed at any point along X; V_f = fibre volume being dyed; V_p = the pore volume of the package; V_s = the external solution volume; $C_s \equiv C_s$ (t) = current dye concentration out of the fibre layer; A ≡ A(X,t) = the mean dye concentration inside the volume of the single fibre; C ≡ C(X,t) = the dye concentration inside the solution located in the fibrous layer; ε - the porosity, equal to the ratio of pore volume to the total package volume; C_0 - the initial dye concentration in the solution.

Moreover, in considering the dependence of sorbate convective transfer rate upon the liquid flow velocity (38), it is necessary to use the following equation [17, 26] to define the alteration of mass-transfer coefficient across the package radius at the boundary conditions defined by (10)

$$\alpha = \alpha_0 \left(\frac{X_0}{X_0 + X} \right)^{\beta}, \qquad (40)$$

where α_0 = the mass-transfer coefficient for the package layer adjoining the spindle.

According to the theoretical analysis of the regularities of convective transfer of substance from the solution to the single fibre surface in the package [27-30], index β in formula (38) is equal to 1/3 and the coefficient can be calculated analytically proceeding from the liquid flow velocity U_0 and from either the liquid specific expenditure or the package hydraulic resistance.

The application of the described procedures of the kinetic curve restoration through the use of moments makes it possible to calculate the dyeing kinetics in the chosen points of the fibrous material package at different temperatures and bath circulation velocities [31, 32]. These procedures demand special algorithms and high computer time expenditures, both of which are sometimes unacceptable.

That is why to analyse the general regularities of the package dyeing process and to carry out engineering calculations, it is possible to propose some formulas for the first dye sorption rate distribution moments $M_1(X)$ which characterise both separated package sections (when the circulation proceeds from the centre to the periphery)

$$\frac{M_1(X)}{m_0} = \frac{1}{1+\frac{\lambda_V}{\Gamma}} \frac{1}{(\nu+1)(\nu+3)} +$$

$$+\frac{1}{(\nu+1)Bi_0}\left[\left(1+\frac{X}{X_0}\right)^{\beta} - \frac{1}{1+\frac{\lambda_V}{\Gamma}} \frac{2}{\beta+2} \frac{X_0}{L} \frac{\left(1+\frac{L}{X_0}\right)^{\beta+2} - 1}{2+\frac{L}{X_0}}\right] +$$

$$+\frac{D\Gamma L}{R^2 2U_0}\left[\frac{X}{L}\left(2+\frac{X}{X_0}\right) - \left(1+\frac{L}{2X_0}\right) \frac{\left(\Gamma+\frac{1-\varepsilon}{\varepsilon}\right)/\lambda_V}{1+\frac{\Gamma}{\lambda_V}}\right]\left(\frac{1-\varepsilon}{\varepsilon} - \frac{1}{\lambda_V}\right)$$

(41)

and the total package

$$\frac{\overline{M}_1}{m_0} = \frac{2}{L(L+2X_0)} \int_0^L \frac{M_1(X)}{m_0}(X + X_0)dX =$$

$$\frac{1}{1+\frac{\lambda_V}{\Gamma}} \left[\frac{1}{(\nu+1)(\nu+3)} + \frac{1}{(\nu+1)Bi_0} \frac{2}{\beta+2} \frac{X_0}{L} \frac{\left(1+\frac{L}{X_0}\right)^{\beta+2} - 1}{2+\frac{L}{X_0}} \right] \pm$$

$$\pm \frac{D\Gamma L}{R^2 U_0} \left[\left(1+\frac{L}{2X_0}\right) \frac{\lambda_V + \frac{1-\varepsilon}{\varepsilon}}{\lambda_V + \Gamma} \right] \left(\frac{1-\varepsilon}{\varepsilon} - \frac{1}{\lambda_V} \right)$$

(42)

Here the sign "+" corresponds to the circulation of the solution from the package centre to the periphery while sign "-" indicates the circulation in the opposite direction.

The results obtained give the opportunity to formulate the dimensionless criterion for the prediction of the levelness of dyeing of a cylindrical fibrous package. This criterion is equal to the ratio of the remainder of the first two moments in the central and peripheral package layers $\Delta M = M_1(0) - M_1(L)$ to the mean moment $\overline{M}_1$ for the package volume $\overline{M}_1$:

$$\chi = \Delta M_1 / \overline{M}_1 \qquad (43)$$

The values of ΔM for both circulation directions are defined by the following correlation

$$\frac{\Delta M_1}{m_0} = \frac{1}{(\nu+1)Bi_0} \left[\left(1+\frac{L}{X_0}\right)^{\beta} - 1 \right] \pm$$

$$\pm \frac{D\Gamma L}{R^2 U_0} \left(1+\frac{L}{2X_0}\right) \left(\frac{1-\varepsilon}{\varepsilon} - \frac{1}{\lambda_V} \right) \qquad (44)$$

It is reasonable to emphasise that geometrically the mentioned moment remainder displays the area between the sorption curves in the centre and periphery of the package and, consequently, characterises the initial dying unlevelness. As the analysis of the formula (23) revealed, there is a close correlation between the mean volume moment of a package and the rate of the dyeing process. All these facts lead to the conclusion that the high values of the proposed criterion correspond to a small dyeing time and high initial unlevelness of textile package dyeings. On the contrary, the small values of the criterion cited correspond to a small risk of uneven dyeing but reduce the rate of the process.

Thus, formulas (42) to (44) enable us to predict the quality and the rate of package dyeing according to the influence of such technological factors as dyeing solution flow velocity, porosity, and package geometry as well as dye sorption properties and dye diffusion in the fibre. The legitimacy of the practical application of the theoretical approach developed for the estimation of the unlevelness of textile package dyeing may be proved by several examples. First, the function of χ-parameter, calculated on the basis of dye solution circulation rate [32] (Fig.1), qualitatively agrees with the experimental data on cotton yarn dyeing with vat dyes [33] (Fig.2). Second, the correlation between the designed parameter and experimental dyeing unlevelness indices, shown in Fig.3., is close to linear [34, 35]. The experimental data correspond to the semi-industrial conditions of high-temperature dyeing of polyester textured threads in the presence of different surface-active agents in a circulation type apparatus. The technical characteristics of the apparatus, the package geometric dimensions, as well as the diffusion coefficients and the partition coefficients of the disperse dye in the film (obtained in the laboratory experiments) were used when calculating the theoretical parameter χ.

The consecutive application of formulas (42)-(44) to calculate the rate of the process and the dyeing unlevelness enables us to define the optimal composition of the dyeing solution and to work out the optimal technological conditions for high-speed dyeing of polyester textured threads [35]. The method created is grounded on the introduction of the restrictions on process duration (according to the value of the parameter $\overline{\mu}_1 = (M_1 / m_0)R^2 / D$), dyeing levelness (according to the parameter χ), dyeing saturation (according to the partition coefficient G). The graphic variant of the definition of the most suitable surfactants is presented in Fig.4. A field confined between three

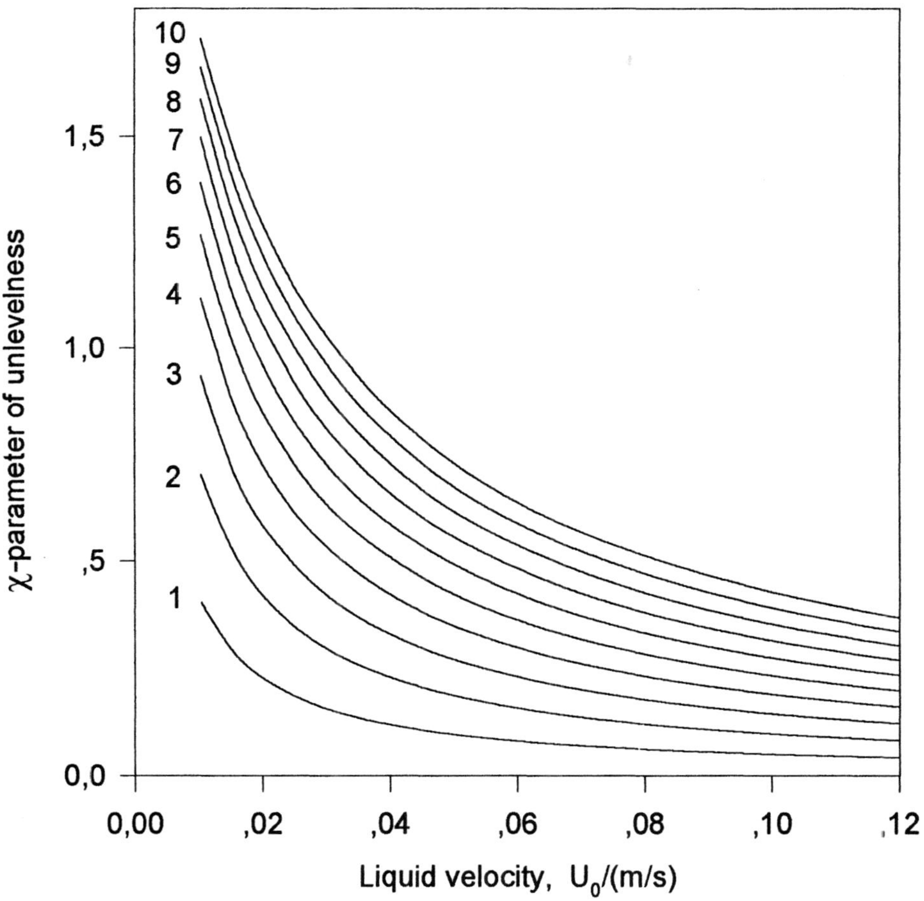

Fig.1. χ-parameter of unlevelness versus mean liquid velocity U_0 for λ_V=12.42, Γ=50, R=$1.5\cdot10^{-5}$ m, ε=0.71, X_0=0.035 m, L=0.0825 m and various diffusion coefficients:

No.	$D/10^{-14}$ m^2/s	No.	$D/10^{-14}$ m^2/s
1	1	6	6
2	2	7	7
3	3	8	8
4	4	9	9
5	5	10	10

Ordinate = χ-parameter of unlevelness

Abscissa = Liquid velocity U_0/(m/s)

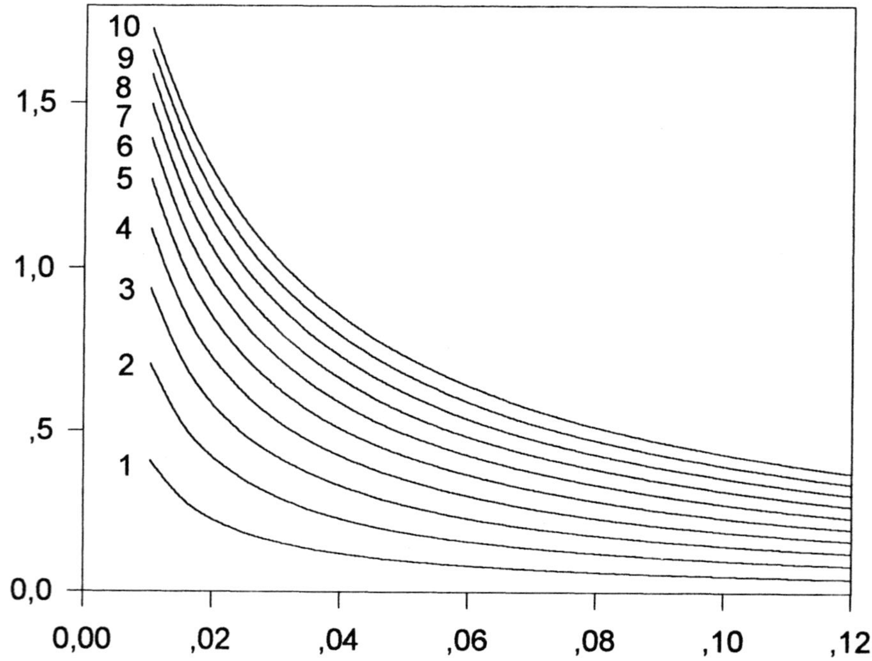

Fig.2. Unlevelness index versus mean liquid velocity for cotton yarn dyeing with the following vat dyes: 1 - isoviolantron; 2 - vat bright green; 3 - vat yellow.

Ordinate = Unlevelness index, %
Abscissa = Liquid velocity, cm/s

curves is an optimal region of coefficients D and G for disperse blue, taking into account, that exhaustion is finished at the stage of 95% degree of equilibrium sorption, practical coefficient of unlevelness does not exceed 4%, and colour yield is not less than 4.5%.

Analysis shows that formula (44) represents the sum of two variables: One which defines the difference between the dyeing rate of the external and internal package layers. This difference is connected with the regularities of the external transfer kinetics of dye to the fibre surface in these layers. The second which defines the difference

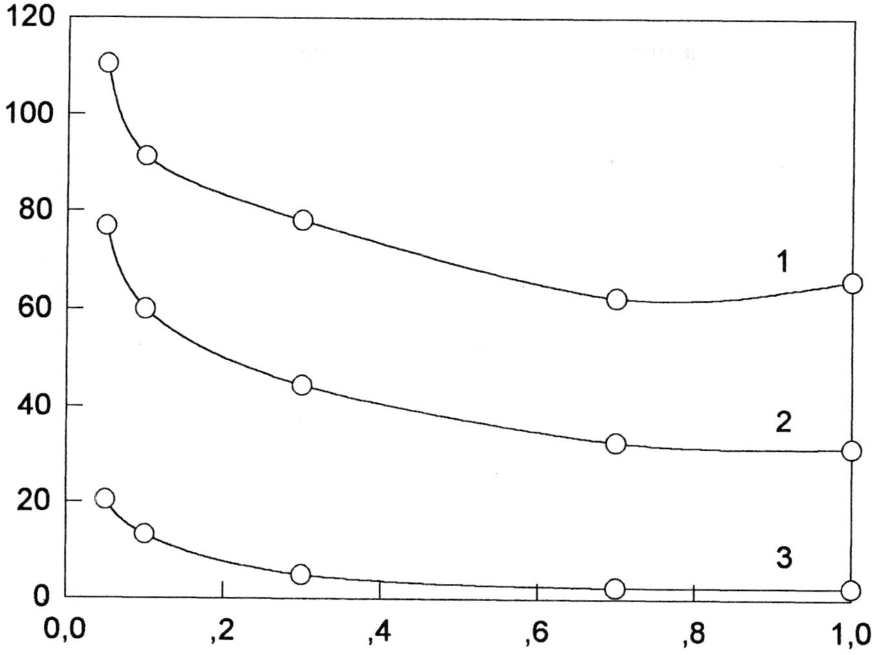

Fig.3. The correlation between real dyeing unlevelness K and the designed parameter χ in the process of dyeing of polyester thread in the presence of different surface active auxiliaries: 1 - without additives; 2-5 - compounds of general formula

$$C_{17}H_{33}COO-CH_2-CHOH-CHO(CH_2CH_2O)_n(CH_2CHOCH_3)_mH;$$ 2 - n=13, m=0; 3 - n=14, m=0; 4 - n=15, m=0; 5 - n=16, m=0.

Ordinate = χ
Abscissa = K, %

connected with the exhaustion of the solution in the course of its flowing inside the package. The calculated χ-parameter dependence upon dyeing solution circulation speed (having been performed in conformity with the real conditions of wool package dyeing at 100 °C [32] and polyester fibre dyeing at 135 °C [35]) revealed that the contribution of the first term in the total moment remainder value is about 1% or even less. So it is obvious that in real dyeing conditions, the dyeing solution exhaustion along the liquid flow direction inside the fibrous material package is the main cause of unlevel dyeing.

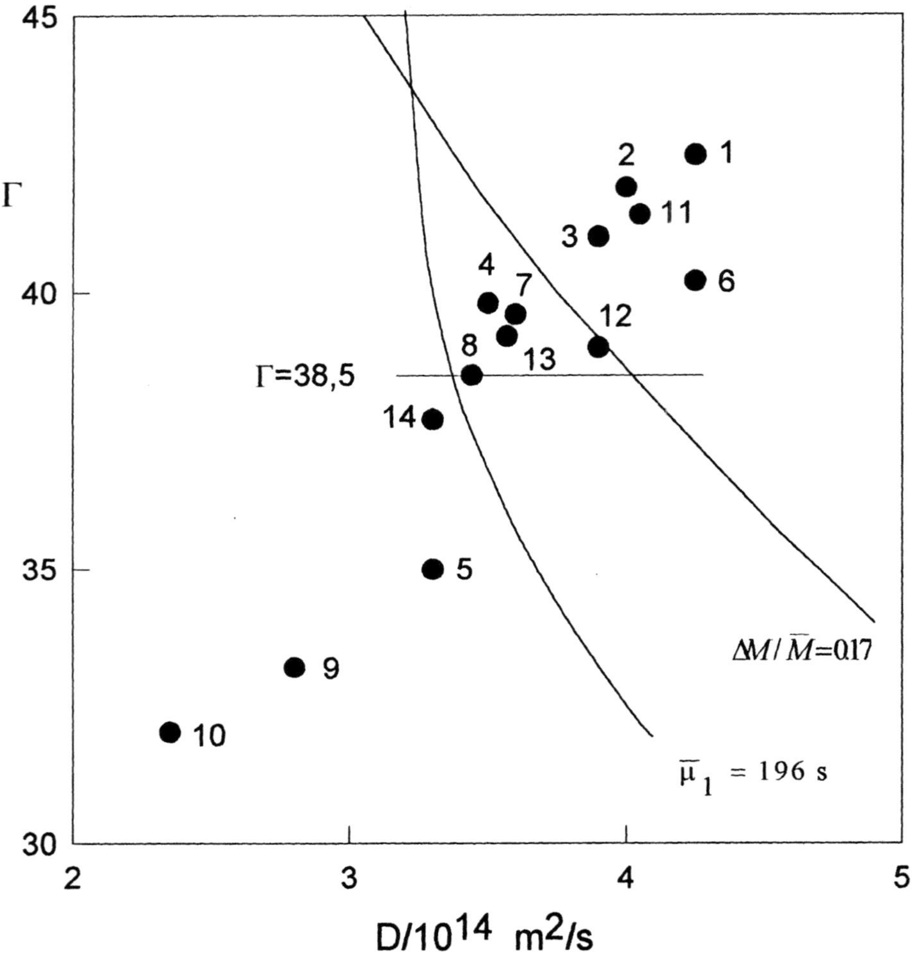

Fig.4. Prognostic diagram for dyeing of polyester thread according to the sorption and diffusion properties of disperse dye in the presence of different surface active auxiliaries: 1 - without additives; 2-10 - compounds of the general formula $C_{17}H_{33}COO-CH_2-CHOH-CHO(CH_2CH_2O)_n(CH_2CHOCH_3)_mH$; 2 - n=13, m=0; 3 - n=14, m=0; 4 - n=15, m=0; 5 - n=16, m=0; 6 - n=8, m=6; 7 - n=12, m=3; 8 - n=13, m=2; 9 - n=13, m=3; 10 - n=13, m=3; 11 - sulfosid 31; 12 - proxanal CL-3; 13 - eganal CME; 14 - karbanox OK-5.

Ordinate = Γ
Abscissa = $D/10^{14}$ m^2/s

The analysis of the formula (42), which defines the general rate of the process, leads to the same conclusion.

By means of disseminating the developed textile packing dyeing process description on the reactive dye sorption processes the following formula was obtained for the zero moment of dye fixation curve $M_{f,0}$:

$$\frac{M_{f,0}}{C_0} = \frac{\Psi \exp\left(-\frac{1-\varepsilon}{\varepsilon}\Psi X\right)}{1-\exp\left(-\frac{1-\varepsilon}{\varepsilon}\Psi L\right)}\frac{V_s}{V_p} + \frac{\varepsilon}{1-\varepsilon}$$

$$\Psi = 2\left(\frac{D\Gamma L}{R^2 U_0}\right)\frac{1}{\frac{I_0(b)}{b I_1(b)} + \frac{1}{Bi}} \qquad (45)$$

This variable, as well as in the formula (37), corresponds to the finite reactive dye distribution in the fibre package. In the procedures, the plane textile material layer was considered with the solution circulating through it in one direction during the whole dyeing process. In the addition, to simplify the calculations, the hydrolysis of dye in the solution was not taken into account.

The analysis of the formula and calculated data obtained demonstrate strong influence of both the circulation speed (Fig.5) and covalent fixation rate constant (Fig.6) on the dye distribution levelness throughout the package volume. That is why in reality (when the intensity of the solution flow through the package remains constant and depends on the dyeing apparatus construction), the dyeing levelness may be regulated by means of changing dye reactivity.

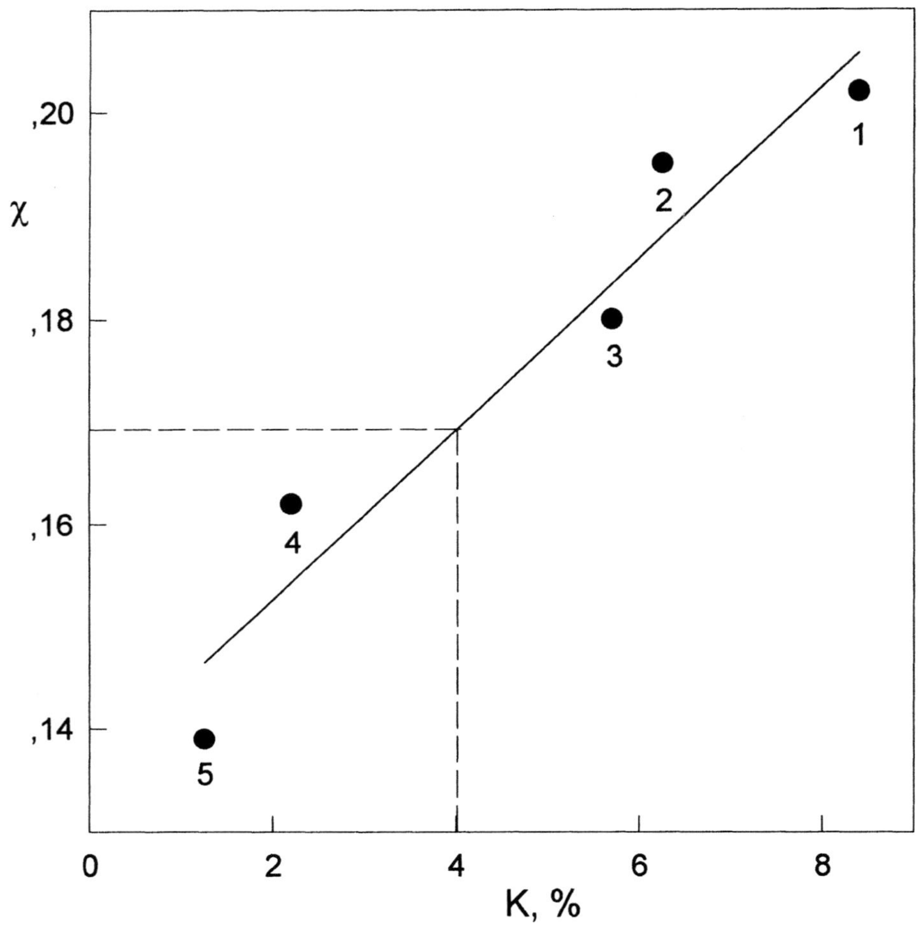

Fig.5. Final distribution of the dimensionless concentration of reactive dye $F_\infty/C_0=M_0/C_0$ along the space coordinate of the package for $k_f=10^{-3}$ s^{-1}, $\Gamma=228$, $D=10^{-13}$ m^2/s, $R=10^{-5}$ m, $\lambda_V=12$, $\varepsilon=0.8$ and different values of mean liquid velocity U_0(cm/s): 1 - 4.9; 2 - 2.5; 3 - 1.6; 4 - 0.4; 5 - 0.1.

Ordinate = M_0/C_0
Abscissa = Space coordinate of the package, m

The examples of application of kinetic curve moment method cited above describe the regularities of dyeing processes. At present, some precedents exist for applying the method of moments to the solution of the problems of interdependent heat- and mass-transfer [36-39] and to the solution of nonlinear diffusion problems provided that the transfer coefficients are the space co-ordinate functions [40]. In the recent publications

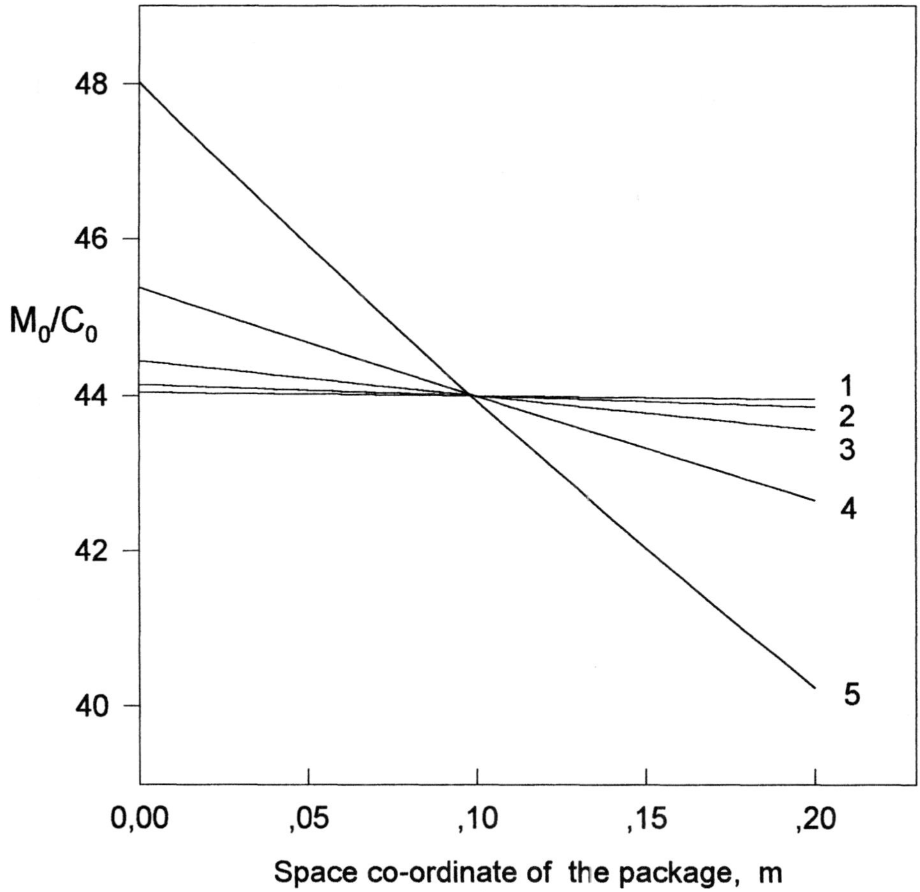

Fig.6. Final distribution of the dimensionless concentration of reactive dye $F_\infty/C_0=M_0/C_0$ along the space co-ordinate of the package for $\Gamma=228$, $D=10^{-13}$ m^2/s, $R=10^{-5}$ m, $\lambda_V=12$, $\varepsilon=0.8$, $U=0.01$ m/s and different values of the constant of covalent fixation rate k_f/s: 1 - 10^{-5}; 2 - $10^{-4.5}$; 3 - 10^{-4}; 4 - $10^{-3.5}$; 5 - 10^{-3}.

Ordinate = M_0/C_0
Abscissa = Space coordinate of the package, m

[41, 42] well-known and new possibilities for the application of the method of moments are systematised. It is natural that the method is also being used to solve problems in chemical kinetics, for example to define the rate constants of the consecutive reaction of vat dye fixation [43].

The results obtained and presented in this article give the hope that, in future, the method of moments will be diffused more widely and will be developed and applied to further the analysis and prediction of complex kinetic process regularities.

References

1. N.N. Tunitskiy, V.A. Kaminskiy, and S.F. Timashev, The Methods of Physical Chemistry Kinetics (Moscow: Khimiya, 1972) 14, 70 (in Russian).

2. Yu.A. Kokotov, P.P.Zolotaryov, G.E.El'kin, Theoretical Bases for Ion-Exchange: Complex Systems of Ion-Exchange (Leningrad: Khimiya, 1986) 135, 248 (in Russian).

3. P.P. Zolotaryov, Yu.M. Popkov and V.V.Ugrozov, J. Physical Chemistry, 62 (10) (1988) 2702-2708 (in Russian).

4. M. Byulov A.M. Voloshchuk, I. Kerger, and V.I. Ulin, Book of Papers: Adsorption and Porosity. The Fifth Conference Devoted to the Theoretical Problems of Adsorption (Moscow: Nauka, 1983) 129-136 (in Russian).

5. F.Yu. Telegin and B.N. Mel'nikov, J. Physical Chemistry, 54 (8) (1980) 2060-2062 (in Russian).

6. B.N. Devyatov and N.D. Demidenko, Theory and Methods for the Analysis of Controlled Distributed Processes (Novosibirsk: Nauka, 1983) (in Russian).

7. M.A. Belov and T.T. Tsyrulis, Assymptotic Methods of Inverse Integral Transformations (Riga: Zinatne, 1985) 17 (in Russian).

8. M.G. Kendall and A. Stuart, Disrtibution Theory (Moscow: Nauka, Glavnaya redaktsiya fiziko-matematicheskoi literatury, 1966) 83 (Russian Edition).

9. F.Yu. Telegin, A.V. Shormanov and A.I. Bashkirov, J. Phys. Chem., 66 (5) (1992) 1309-1314 (in Russian).

10. F.Yu.Telegin, in: B.N. Mel'nikov, A.P. Moryganov and Yu.A. Kalinnikov, Theory and Practice of Intensive Fixation of Dyes on Textile Materials, (Moscow: Legprombytizdat, 1987) 4-14 (in Russian).

11. M.A. Lavrentyev and B.V.Shabat, The Methods of the Theory of Functions of Complex Variable (Moscow: Nauka, Glavnaya redaktsiya fiziko-matematicheskoi literatury, 1987) 459 (in Russian).

12. G.A. Baker, Jr. and P. Graves-Morris, Padé-Approximants (Moscow: Mir, 1986) (Russian Edition).

13. F.Yu. Telegin, A.V. Shormanov and S.V.Sazhin, Papers of the conference of young scientists of Chem. Dept. of Moscow State University, Moscow, Jan. 26-28, 1988: Part 2. (Moscow) (1988) 179-182 (in Russian).

14. Wilkinson and Reinsh, Handbook for Algorithms in ALGOL: Linear Algebra (Moscow: Mashinostroenie, 1976) 316 (Russian Edition).

15. F.Yu. Telegin, A.V. Shormanov and A.I. Bashkirov, J. Physical Chemistry, 66 (5) (1992) 1315-1320 (in Russian).

16. F.Yu. Telegin and B.N. Mel'nikov, Izv. Vuzov. Khimiya i Khim. Tekhnol., 28 (12) (1985) 101-104 (in Russian).

17. B.N. Mel'nikov and F.Yu. Telegin, Physical Chemical Bases for Control of Dyeing Processes (Moscow: Legprombytizdat, 1989) (in Russian).

18. F.Yu. Telegin, B.N. Mel'nikov and I.B. Blinicheva, J. Physical Chemistry, 56 (1) (1982) 201-203 (in Russian).

19. Crank J, The Mathematics of Diffusion, 2-nd ed. (Oxford: Clarendon Press, 1975).

20. F.Yu. Telegin, A.I. Bashkirov and A.V. Shormanov, J. Physical Chemistry, 66 (6) (1992) 1549-1554 (in Russian).

21. G.E. Krichevskiy, The Physicochemical Bases for the Use of Reactive Dyes (Moscow: Legkaya Industria, 1977) (in Russian).

22. R.H. Peters, Textile chemistry. Vol.3: The Physical Chemistry of Dyeing (Amsterdam-Oxford-Ney York: Elsevier Sci. Publ. Co., 1975) 771-819.

23. F.Yu. Telegin, B.N. Mel'nikov and I.B. Blinicheva, Izv.Vuzov. Tekhnol. tekst. prom. (4) (1982) 56-59 (in Russian).

24. B.N. Mel'nikov et al., The Progress in Textile Chemistry (Moscow: Legprombytizdat, 1988) (in Russian).

25. I.I. Zaslavskiy, The Bases for Theory of Dyeing with Ionic Dyes (Moscow: Legprombytizdat, 1989) (in Russian).

26. F.Yu. Telegin, A.V. Shormanov and B.N.Mel'nikov, Izv. Vuzov. Khimiya i Khim. Tekhnol., 30 (12) (1987) 91-95 (in Russian).

27. F.Yu. Telegin et al., Izv. Vuzov. Khimiya i Khim. Tekhnol.,30 (6) (1987) 87-91 (in Russian).

28. F.Yu. Telegin et al., Izv. Vuzov. Khimiya i Khim. Tekhnol.30 (6) (1987) 92-96 (in Russian).

29. F.Yu. Telegin, V.B. Kuznetsov and A.V. Shormanov, Papers of the Conference of Young Scientists, Chemical Dept. of Moscow State University, Moscow, Jan. 26-28, 1988: Part 2. (Moscow, 1988) 202-205 (in Russian).

30. V.B. Kuznetsov, F.Yu. Telegin and B.N.Mel'nikov, Rapid Washing of Fabrics after Dyeing and Printing, in Reviews of Cotton Industry, (5) (Moscow: CNIITEILegprom, 1990) (in Russian).

31. A.V. Shormanov, F.Yu. Telegin and B.N. Mel'nikov, Izv. Vuzov., Tekhnol. Tekst. Prom., (2) (1989) 56-60 (in Russian).

32. A.V. Shormanov, Developing the Highly Effective Method for Textile Material Dyeing in the Batch Dyeing Process (Dissertation for the degree of Candidate of technical sciences. Ivanovo, 1989) (in Russian).

33. V.F. Androsov et al., Dyeing of Yarn in Packages (Moscow: Legkaya Industriya, 1974) (in Russian).

34. S.G. Belokurov et al., Izv. Vuzov. Tekhnol. Tekst. Prom. (5) (1990) 65-69 (in Russian).

35. S.G. Belokurov, Theoretical Bases and the Newly Developed Process for Package Dyeing of Polyester Threads. (Dissertation for the degree of Candidate of technical sciences. Ivanovo, 1991) (in Russian).

36. F.Yu.Telegin et al., Izv. Vuzov. Tekhnol. Tekst. Prom. (6) (1985) 50-52 (in Russian).

37. F.Yu. Telegin et al., J. Physical Chemistry, 60 (8) (1986) 2039- 2042.

38. V.V. Kozlov, Intensification of Textile Material Impregnation Process by the Use of Vapor Pretreatment (Dissertation for the degree of Candidate of technical sciences. Ivanovo, 1984) (in Russian).

39. M.N. Gerasimov, The Application of Vapor Treatment of Textile Materials for the Improvement of Finishing Processes (Dissertation for the degree of Doctor of technical sciences, St-Petersburg, 1991) (in Russian).

40. N.N. Tunitskiy, Book of Papers: Adsorption and Porosity. The Forth Conference Devoted to the Theoretical Problems of Adsorption (Moscow: Nauka, 1976) 285-290, 323-324 (in Russian).

41. A.I. Volzhinskiy and V.A.Konstantinov, Regeneration of Ionites. The Theory of the Process and Calculations of the Apparatus (Leningrad: Khimiya, 1990) (in Russian).

42. A.M. Azizov, Analysis of the Technological Processes. Parametric and Non-linear Phenomena (Leningrad: Khimiya, 1992) (in Russian).

43. O.I.Odintsova et al., Izv. Vuzov. Tekhnol. Tekst. Prom. (6) (1990) 58-60 (in Russian).

Department of Chemical Technology of Fibrous Materials, Ivanovo State Academy of Chemical Technology, Engels av., 7, Ivanovo, 153460, Russia.

THE THEORETICAL BASES FOR THE FORMATION OF AGGRESSIVE-RESISTANT ELASTOMERIC COVERS ON ARTIFICIAL LEATHERS

L.N.Mizerovskiy

The theoretical substantiation of the methods of the formation of aggressive-resistant elastomeric coverings on artificial leathers has been proposed. The possibilities for the practical realization of the discussed theoretical approaches have been considered.

At present, the production of protective clothes is one of the most important practical applications of artificial leather (AL). In many cases, the requirements for the properties of such materials include high resistance to the influence of organic liquids within a large temperature range ($+30^0C \div -50^0C$) provided that the elasticity of the materials is preserved. This condition for elasticity leads to the necessity of the application of vulcanizates of different elastomers (which, unlike amorphous-crystal polymers, can dissolve many organic liquids during swelling [1]) as polymeric coverings of AL

As a result, the problem arises on how to produce aggressive-resistant elastomeric AL coverings. This problem consists of two parts. First, the solubility of a liquid in a vulcanizate must be limited in order not to worsen the set of physico-mechanical indices of a polymeric covering and of the AL as a whole. Second, the solvent penetration through AL must be delayed as long as possible.

In spite of the closeness of these two parts from the physicochemical point of view, they are not identical and must be considered separately. The present article is devoted to the discussion of the first part and (in contrast with other monographs [1-4]) places its emphasis on the analysis of different ways of providing the solution.

According to the Russian standard (GOST 9.071-76) which establishes the requirements for the aggressive resistance of resins (applied to goods which are used in liquid aggressive media provided that the resins are not subjected to tension[1]), the permitted change in resin mass within the groups of resin resistance 1, 2, and 3 is equal to (%): -3 to +12; -5 to +30, and -50 to +30, respectively.

As the contact of physico-active liquid with resin can be accompanied by the outflow of liquid-soluble ingredients of the resin, the value of measurable mass accretion will be lower than the real solubility of the liquid, with the difference being equal to the mass of washed out ingredients. That is why the permitted level of the solubility of liquids $Q_{max.}$ in resins of different aggressive resistance is accepted to be 15, 35, and 80% for the groups 1, 2, and 3, respectively.

Undoubtedly these values are conditional, and for the polymeric covering of each concrete AL, the value of permitted equilibrium solubility of physico-active liquid in the polymer used must be stipulated according to the actual requirements for the material being produced[2**]. In the case under discussion, the number of $Q_{max.}$ is of paramount importance because it determines theoretically the possible ways of regulating the aggressive-resistance of AL while simultaneously retaining all of AL's quality indices at the required level.

Let us examine from the considered positions the influence of elastomer structure; the vulcanization reticule's density; the nature and concentration of a filler; and the character of the distribution of the polymer mixture's components on the equilibrium solubility of physico-active liquids in vulcanizates.

The Chemical Composition Of Elastomers.

Obviously, this parameter is the most important one, and the low thermodynamic affinity of a liquid towards an elastomer is the necessary and sufficient condition for the formation of an aggressive-resistant composition using this liquid as a main component. It is quite clear that if the equilibrium solubility of the liquid in the polymer $Q \leq 15\%$,

---

[1] To say concisely: the state of AL polymeric covering in the conditions under which the protective cloth is used can not be regarded as non-tensed, because at definite places a covering is subjected to multiple bend deformations but these deformations can be disregarded due to the insignificance of their amplitude and frequency.

[2] The adopted technical requirements for AL used in protective cloths anticipate an increase in material mass (due to the sorption of a physico-active liquid) of not more than for 5-10% during 24 hours.

the vulcanizates formed on the base of this polymer belong to group 1 ; if Q ≤ 80%, the vulcanizates can be attributed at least to group 3; if Q > 80%, the prediction of the achievable level of vulcanizate aggressive resistance is impossible without acquiring additional data.

In the scientific literature, information about the equilibrium solubility of organic liquids in elastomers is absent. Nevertheless, the value of this solubility can be estimated with sufficient accuracy using the following expression [5]:

$$Q = \left(\frac{0.476}{\chi_1 - 0.3} \right)^2 \qquad (1)$$

which is the approximate solution of the well-known Flory-Rener's equation presented in the form of the equality:

$$(1+Q)^2 \ln\left(1+\frac{1}{Q}\right) - (1+Q) = \chi_1 + \frac{V_1 \rho_2}{\overline{M_c}} \left[(1+Q)^{\frac{5}{3}} - \frac{1}{2}(1+Q) \right]$$

where Q = the equilibrium volumetric degree of vulcanizate swelling; V_1 = the liquid mole volume; $\overline{M_c}$ = the molecular mass of the part of the chain located between the knots of the polymer reticule; χ_1 = the interaction parameter of Flory-Higgins (the condition that must be fulfilled is that the second term in the right part of the equation is much less than χ_1). The physical sense of this condition consists in the fact that a definite critical swelling degree $Q_{crit.}$ (beginning from which the realization of the inequality $Q < Q_{crit.}$ means that the vulcanizate acts as a linear polymer insoluble in the liquid [5, 6]) corresponds to each $\overline{M_c}$ value.

On the other hand, as far as the limiting values of Q (which give the possibility to attribute a polymer and, consequently, a vulcanizate formed on its base to a definite aggressive resistance group) are known, the values of $\chi_{crit.}$ can be calculated using the equation (1), and the condition of the aggressive resistance can be expressed by the simple inequality $\chi_1 >> \chi_{crit.}$ For the first, second, and third groups. The value of $\chi_{crit.}$ ($\chi'_{crit.}$, $\chi''_{crit.}$, $\chi'''_{crit.}$) equals 1.53, 1.10, and 0.83, respectively. The values of the parameter χ_1 for many "polymer - liquid" pairs are combined in the works [7, 8] while the data on the swelling of different vulcanizates are given in [3, 9].

All the above said give the utmost evidence of the expediency of the attempts to calculate χ_1 using only the physicochemical characteristics of a pair "polymer - liquid". Usually the parameter χ_1 is represented as a sum

$$\chi_1 = \chi_s + \frac{V_1}{RT}(\delta_1 - \delta_2)^2 \qquad (2)$$

where χ_s = the entropy component of the parameter and where δ_1 and δ_2 = the parameters of solubility of a liquid and a polymer, respectively.

Let us estimate the necessary accuracy of χ_1 calculations in order to make an error of the prediction of the equilibrium solubility of a liquid in a polymer (or the degree of polymer swelling) not higher than, for example, ± 10%.

As comes from equation (1), in the region of $\chi_1 = 0.8$, this error is not less than ± 0.03 whereas in the region of $\chi_1 = 1.5$ not less than ± 0.06. That is why, if the contributions of both terms of the equation (2) in the summary calculation error of χ_1 are equal, the error of the estimation of χ_s does not exceed ± 0.015 ÷ 0.030, and the error of the evaluation of (δ_1 - δ_2) is less than 40 ± 60 $(J/m^3)^{1/2}$ at 298 K with the mole volume of a liquid being equal to 10^{-4} m^3/mol.

Having compared these values with the temperature coefficient of the solubility of usual liquids (± 30 $(J/m^3)^{1/2}$ [10]) and with the error of δ_1 and δ_2 calculations according to the rule of additivity (~400 - 600 $(J/m^3)^{1/2}$ [11]), it is easy to come to the conclusion that the exact prediction of Q presupposes the use of extremely precise values of the parameters of solubility of both components: a liquid and a polymer.

Such δ values for more than 100 pure liquids at 298 K are combined in manuscript [12]. The accuracy of the parameter of polymer solubility is usually unknown [13]. That is why it is assumed (at least, at present) that the direct experimental definition of the equilibrium solubility of liquids in elastomers within a wide temperature range is the most reliable way to obtain that information which is needed to predict the stability of vulcanizates in physico-active media.

Density Of The Vulcanizate's Reticule.

The conviction exists [1-4] that it is possible to regulate the resistance of resins against the influence of physico-active media by means of altering the nature and density of the vulcanization reticule. In reality, this is not exactly true. As was already noted, a vulcanizate acts with respect to a liquid as a three-dimensional polymer only if $Q > Q_{crit.}$. That is why the possibility to regulate the level of vulcanizate swelling by altering $\overline{M}_c$ arises only when the value of $\overline{M}_c$ (as defined by the requirements for the deformation resistance properties of vulcanizates) is less than those resulting from the requirements for the limiting value of Q.

In conformity with [6], the correlation between $Q_{crit.}$ and $\overline{M}_c$ is described by the empirical equation:

$$Q_{crit.} \approx 0.253\sqrt[4]{\overline{M}_c} - 1.342.$$

The calculations made using this equation give the following values of $\overline{M}_c$ and $Q_{crit.}$:

$\overline{M}_c \cdot 10^{-3}$, g/mol	$Q_{crit.}$
20.0	1.67
10.0	1.19
5.0	0.789
2.5	0.45
1.25	0.16

Because the value of an $\overline{M}_c$ parameter of an AL polymer covering lies in the region of ~10^{-4} g/mol [14], it is clear that a decrease in the degree of polymer swelling in pysico-active media even down to the level of the requirements for the polymers of the third group (due to the transformation of the elastomer into its vulcanizate) is impossible.

A carbo-chain elastomer vulcanizate, containing bonds differing from carbon-carbon bonds, can be considered as a copolymer whose thermodynamic affinity towards a definite liquid is not equal to the same parameter characterizing the initial elastomer. Nevertheless, if the vulcanization reticule is relatively sparse, the mentioned effect also can not be substantial. In any case, the data presented in [9] show that in the region where $\overline{M_c}$ = (10-20)·10^3, the differences between the degree of swelling of the radiation and sulfur vulcanizates in the media possessing small affinity to elastomers are not observed.

The Nature And Concentration Of A Filler

The introduction of fillers into elastomers is a well-known device for decreasing the degree of swelling of vulcanizates in physico-active media. This device is based on three effects: first, the diminution of a portion of sorption-active component; second, the decrease in sorption capacity of polymeric matrix due to the increase of the effective mean density of the vulcanization reticule caused by re-vulcanization of the elastomer micro-volumes directly contacting with the particles of active filler; and third, the mutual strengthening of the structural reticules of the filler and the polymer [15, 16].

Attention must be paid to the fact that all these effects become apparent only when the swelling degree of unfilled matrix is measured in hundreds of per cent, and when in comparison with this degree, the sorption of the liquid by the filler itself is negligible [1, 15, 17-19].

In the range of vulcanizate swelling degrees within our research interest (corresponding to such a state of the system "vulcanizate - solvent" in which the vulcanizate displays the properties of a linear insoluble polymer relative to a liquid), all the encountered effects, except the first one, either are not displayed at all or possess extremely low values. Vice versa, the sorption of the solvent by the filler may turn out to be comparable with the filler solubility in the polymer. The data obtained in [20] can serve as a corroboration of the discussed theoretical aspects. According to these data, the sorption capacity of chalk varies from 0.09 to 0.54 g/g, of kaolin - from 0.27 to 0.61 g/g, and of Aerosil - from 1.98 to 5.40 g/g. These figures demonstrate that even if a relatively small part of filler surface is accessible for the liquid inside the vulcanizate, the contribution of the filler to the total sorption capacity of resin can be considerable.

In the range where $Q < Q_{crit.}$, it is expedient to present the mass degree of swelling of filled vulcanizate as a sum of two terms:

$$Q = \alpha Q^{*}_{fil.} + (1-\alpha) Q_{el} \qquad (3)$$

where α and $Q_{fil.}$ = the mass fraction and the equilibrium sorption capacity of the filler in the vulcanizate; $Q_{el.}$ = the mass degree of polymer matrix swelling which can be calculated according to the expression:

$$Q_{el.} = \frac{\rho_1}{\rho_2}\left(\frac{0.476}{\chi_1 - 0.3}\right)^2 \qquad (4)$$

where ρ_1 and ρ_2 respectively = the densities of the liquid being sorbed and of the rubber.

Today it is still impossible to estimate theoretically the value of $Q_{fil.}^*$. However, this parameter can be defined experimentally if one uses the data on the swelling of the filled vulcanizates with a known value of $\overline{M_c}$ in the series of liquids whose behavior obeys equation (4). The values of $Q_{fil.}^*$ for chalk, kaolin, and Aerosil calculated in [20] according to the results for the mass swelling of the radiation vulcanizates SKD, SKEPT -50, SKTV-K, and SKN-40 and of sulfur vulcanizate SKD in different liquids are given in Table 1.

Table 1.

Polymeric matrix	$Q_{fil.}^*$, g/g		
	chalk	kaolin	Aerosil
ÑÊÄ (sulfuric)	0.15±0.08 0.26±0.06	-	-
ÑÊÄ	.04 ± 0.07 0.26 ± 0.06	0.0 ± 016 0.38 ± 0.12	0.19 ± 0.09 3.0 ± 0.5
ÑÊÝÏÒ-50	0.05 ± 0.09 0.30 ± 0.20	0.07 ± 0.08 0.43 ± 0.16	0.48 ± 0.09 3.3 ± 1.0
ÑÊÒÂ-Ê	0.08 ± 0.09 0.29 ± 0.07	0.20 ± 0.15 0.44 ± 0.15	0.03 ± 0.1 3.4
ÑÊÍ-40	0.54 ± 0.30 0.30 ± 0.20	0.31 ± 0.20 0.41 ± 0.14	0.35 ± 0.15 3.6 ± 1.6

Note: Under the lines, the mean values for the sorption of liquids (used in the experiments on the swelling of filled vulcanizates) on the corresponding filler are given.

In spite of the fact that the large scatter in the found values of $Q_{fil.}^{*}$ is observed, their analysis gives the possibility to draw a conclusion which is of great scientific and applied interest.

1. The degree of the accessibility of the filler's surface inside the vulcanizate for the sorption of liquids depends significantly on the nature of the polymer matrix and filler and on the method of vulcanization. In the case under discussion, this degree varies up to ~1% for the system "SKTV-K - Aerosil" and up to ~100% for the systems "SKN-40 - chalk" and "SKN-40 - kaolin".

2. The principle of additivity of the adsorption of non-active (in respect to the polymer matrix of filled vulcanizate) liquids is fulfilled rather well (it is the basic principle of equation (3)). This gives grounds to confirm that when the thermodynamic affinity of a liquid to an elastomer is low, the introduction of any filler into this elastomer can influence the vulcanizate sorption capacity only because of the decrease in the polymer portion in the vulcanizate and only on the condition that the filler's surface is almost completely "blocked" by a polymer which can be easily wetted.

The Vulcanizates Formed On The Base Of Elastomer Mixtures

As is known, there are three possible reasons for the observed deviations of the degrees of equilibrium swelling of the vulcanizates formed by the elastomer mixtures from the additive values: The first reason consists of the changes in the densities of the space reticules of the mixture components in comparison with the reticule densities in the corresponding one-phase systems. The second reason is the existence of the developed inter-phase layer. And the third is the transfer of the system from swelling in a free state to swelling in a tensed-deformed state that is dependent on the character of phase distribution inside the volume of the composition. In publication [21], it has been shown that among these three, only the last reason can play a substantial practical role.

Visual evidence of the influence of polymer mixture phase structure on the solubility of physico-active liquid in this mixture can be obtained if one considers a binary composition as a ball element consisting of a nucleus and an outer cover.

Qualitatively, it is clear that if the cover swells less intensively than the nucleus and possesses sufficient mechanical strength, it would restrict the swelling of the nucleus, even in spite of a certain increase in the cover's sorption capacity due to tension deformation. If, on the contrary, the cover swells more intensively than the nucleus, it would not impede the growth of nucleus volume because the lower the value of Q, the greater is the swelling pressure. As a result, the total swelling of the system would be governed by the rule of additivity.

If the nucleus radius is designated as R_0 and the density of the cover in the non-swollen ball element is h_0, it is possible to show [21] that the following equality will correspond to the condition of mechanical equilibrium in the swollen system:

$$\frac{h_0}{R_0} = \frac{3}{2E_\infty\left(p_0^{-1/2} - \dfrac{Q_c - Q_n}{k}\right)^2} \cdot \frac{(Q_n+1)^2\sqrt[3]{Q_n+1}}{(Q_n+1)^2 - 1} \quad (5)$$

where E_∞ = the conditionally-equilibrium module of the cover's elasticity; Q_n and Q_c = respectively, the equilibrium volumetric swelling degrees of the nucleus and the cover; p_0 = the swelling pressure in the swollen nucleus in the absence of the cover at equilibrium; k = the proportionality coefficient. This equation determines the correlation between the vulcanizate swelling in a free state and in a state of overall compression [21].

If the values of Q_n, Q_c, E_∞, p_0, and k are either known or predefined, it is possible using equation (5) to calculate the value of h_0/R_0 and, subsequently, the ratio of the initial volumes of the nucleus and the cover as well as the value of equilibrium swelling of the system as a whole. Let us analyze an example. Two ball elements exist whose nuclei are made of vulcanizates with the following characteristics [21]: nucleus I: k = 5.97 $MPa^{1/2}$, p_0 = 0.081 Pa, Q_n = 3.0; nucleus II: k = 3.14 $MPa^{1/2}$, p_0 = 0.172 Pa, Q_n = 0.3. Let the cover be made of vulcanizates or thermoplasts with E_∞ lying in the interval of 1.0 - 10 MPa and Q_c = 0.05. The calculations demonstrate that, in this case, the dependencies of the equilibrium degree of composition swelling upon the composition content are described by the curves presented in Fig.1.

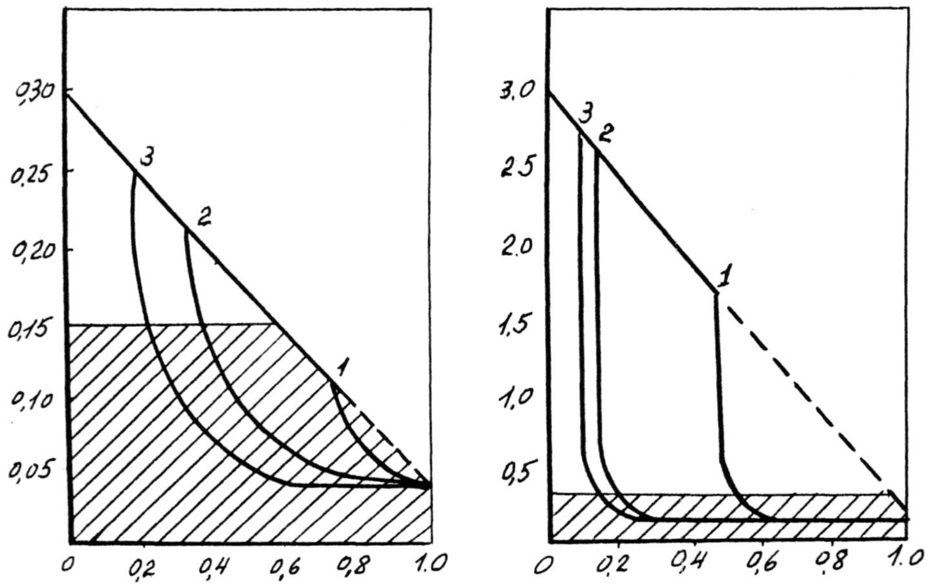

Fig.1. The influence of cover hardness and of nucleus aggressive-resistance on the equilibrium swelling of a model ball element: E_∞ = 1 (1), 5 (2), and 10 (3) MPa. Hatched areas correspond to the ranges of aggressive-resistant compositions of the first (a) and second (b) groups.
X = Volumetric fraction of a cover;
Y = Q (volumetric swelling degree).

It is seen distinctly that the existence of a sufficiently resilient and aggressive-resistant cover causes an abrupt decrease in the equilibrium degree of swelling of the whole composition. The more markedly this effect occurs, the larger is the thermodynamic affinity of the physico-active liquid to the nucleus material. In other words, the formation of such structure of binary polymer mixture (which in its physical meaning corresponds to the model "nucleus - cover") allows in principle the creation of an aggressive-resistant composition –even in that case when one of the mixture's components swells intensely in the given physico-active liquid. The efficiency of the described approach is corroborated by the simulating experiments described in article [22] as well as by the examples of a large increase in resin aggressive resistance owing to the formation of a surface oxidized layer or owing to the co-vulcanization of rubber with thermoplasts [1, 2, 4, 23-26].

Conclusions

The above discussion gives proof of the existence of three possible ways to create an elastomeric composition which would not dissolve a quantity of physico-active liquid exceeding some admissible value and which would possess pre-designed physicochemical indices ($\overline{M_c}$):

1. The application of a single elastomer as a polymer matrix. This elastomer must satisfy the following condition for a given liquid (or liquids): $\chi_1 > \chi'_{crit.}$ ($Q < Q'_{max.}$).

2. The use of such elastomer composition that adsorbs physico-active liquid in conformity with the inequality:

$$\sum_1^n Q_i \alpha_i < Q'_{max.}$$

where Q_i and α_i = sorption capacity and mass fraction of i- component. This component can be represented either by the polymer with a mean ($\chi_1 > \chi''_{crit.}$), small ($\chi_1 > \chi'_{crit.}$), or extremely small ($\chi_1 >> \chi'_{crit.}$) affinity towards the liquid or by an inorganic filler which is easily wetted by the polymer and which is, consequently, characterized by low sorption capacity.

3. The formation of the composition with colloid-chemical structure submitting to the model "nucleus - cover" by means of using the mixtures of the elastomer (χ_2 $\chi'''_{crit.}$) and the linear or three-dimensional high-module polymer ($\chi_1 >> \chi'_{crit.}$). Such a composition can be represented either by the system of mutually penetrating reticules or by the elastomer dispersion in thermoplasts, the redundancy of thermoplasts being not necessary [26, 27].

It is obvious that the first way is the most drastic because it is based on the thermodynamic incompatibility of the polymer and the liquid. However, this way is difficult to realize in practice because very often the requirements for the chemical structure of a polymer, from the point of view of aggressive resistance and physicomechanical properties, appear to be just the opposite (for example, the combination of oil- and freeze-resistance). Economic considerations can also play a notable role in limiting the practical possibilities to apply this approach.

The method of chemical modification of elastomers by means of co-vulcanization with different reactive resins [28, 29] (which has been in use for a long time) as well as the method of the formation of elastomer mixtures with thermoplasts [23, 25] are examples of the third way of regulating the solubility of physico-active liquids in resins. This way, however, is more suitable for creating materials whose elasticity must not be as high as the elasticity of artificial leathers used in protective cloths.

This is why at present, almost all artificial leathers (for protective cloths) possess polymeric coverings formed on the basis of mixtures of filled polymers characterized by low and mean affinity towards physico-active liquids. Due to the introduction of the fillers, not only is the portion of sorption-active matrix in the composition increased, but also (and not less important) a vulcanizate structure is formed which possesses sufficiently high physico-mechanical properties even if each mixture component adsorbs an equilibrium amount of physico-active liquid. Examples of the practical realization of the problem of creating artificial leathers satisfying a complex of different requirements exist in manuscript [30, 31].

References

1. Yu.S. Zuev, The Destruction of Polymers under the Influence of Aggressive Media (Moscow: Khimiya, 1972) (in Russian).

2. Yu.S. Zuev, The Destruction of Elastomers while Using (Moscow: Khimiya, 1980) (in Russian).

3. G.Ya. Voroby'eva, The Chemical Stability of the Polymeric Materials (Moscow: Khimiya, 1981) (in Russian).

4. Yu.S. Zuev and T.G.Degteva, The Stability of Elastomers while Using (Moscow: Khimiya, 1986) (in Russian).

5. N.I. Lytkina, L.N. Mizerovskiy and L.N. Vansyatskaya, Vysokomolekul'arniye Soyedineniya, 25 (4) (1983) 263 (in Russian).

6. L.N. Mizerovskiy, N.I. Lytkina and L.N. Vansyatskaya, Izv.Vuzov. Khimiya i Khim. Tekhnol., 30 (10) (1987) 76.

7. C.I. Sheehan and A.L. Bisio, Rubber Chem. Technology, Rubber Rev., 39 (1) (1966) 149.

8. L.N. Mizerovskiy and T.V. Zaytchikova, The Calculation of the Interaction Flori-Higgins's Parameter Using the Data on the Equilibrium swelling of the Elastomer Vulcanizates (Published in manuscript form No.2344-B91) 52 (in Russian).

9. L.N. Mizerovskiy and L.N. Vansyatskaya, Swelling of the Unfilled Vulcanizates of Several Elastomers in Individual Organic Liquids (Published in manuscript form No.2676-ëï-89) 35 (in Russian).

10. L.N. Mizerovskiy and L.N. Vansyatskaya, Izv.Vuzov. Khimiya i Khim. Tekhnol., 31 (4) (1988) 51 (in Russian).

11. A.A. Askadskiy and Yu.I. Matveev, The Chemical Structure and Physical Properties of Polymers (Moscow: Khimiya, 1983) (in Russian).

12. L.N. Mizerovskiy and L.N. Vansyatskaya, The Parameters of Solubility of Organic Liquids at 298 K (Published in manuscript form No.2410-ëï-88) 13 (in Russian).

13. A.E. Nesterov, TheProperties of Polymer Solutions and Polymer Mixtures (Kiy'ev: Naukova Dumka, 1984) (in Russian).

14. R.N. Khelevin et al., Kozhevenno-obuvnaya Promyshlennost, (1) (1976) 48 (in Russian).

15. G. Krauss, J. Appl. Polymer Sci., 7 (3) (1963) 861.

16. N.N. Lezhnev et al., Kautchuk i Rezina, (2) (1965) 16 (in Russian).

17. L. Millins and N.R. Tobin, J. Appl. Polymer Sci., 9 (1965) 2993.

18. A.S. Lykin et al., Kautchuk i Rezina, (7) (1982) 8 (in Russian).

19. E.A. Anphimova, A.S. Lykin and B.I. Anphimov, Vysokomolekulyarniye Soyedineniya A, 34 (2) (1982) 414 (in Russian).

20. L.N. Mizerovskiy et al., Kautchuk i Rezina, (3) (1987) 17 (in Russian).

21. L.N. Mizerovskiy, Vysokomolekulyarniye Soyedineniya A, 30 (12) (1988) 2473 (in Russian).

22. L.N. Mizerovskiy, L.N. Vansy'atskay'a and G.I. Smurova, Vysokomolekulyarniye Soyedineniya Á, 30 (12) (1988) 883 (in Russian).

23. J.J. Danic and D.C. Edwards, Rubber World, 149 (5) (1964) 36.

24. A.J. Coran and R. Patel, Elastomerics, 122 (5) (1980) 28, 29.

25. A.J. Coran, R. Patel and D. Williams, Rubber Chem and Technol., 55 (4) (1982) 1063.

26. E.E. Piskunova, V.N. Kuleznev and V.V. Korovkin, Novel Developments in the Production of Artificial Leathers and Film Materials (Moscow, 1983) 105 (in Russian).

27. S.A. Komarov et al., Colloid J., (1) (1986) 152 (in Russian).

28. A.G. Shvartz and B.N. Dinsburg, The combination of Rubbers with Plastics and Synthetic Resins (Moscow: Khimiya, 1972) 224 (in Russian).

29. A.A. Dontsov et al. Rubber and Resin, (3) (1985) 9 (in Russian).

29. S.V. Uskova, V.P. Kitaev and L.N. Mizerovskiy, The Development of the Technology for Producing the Artificial Leather resistant to Pesticides (Published in manuscript form No.1812-ëï-86) 13 (in Russian).

30. S.V. Uskova et al., The Use of Micro-computers for the Calculations of the Compositions of Binary Mixtures resistant to Physico-active Media (Published in manuscript form No.2341-ëï-88) 13 (in Russian).

The Institute of Chemistry of Non-Aqueous Solutions
of the Russian Academy of Sciences, Ivanovo

THE MECHANISM OF THE INTENSIFICATION OF WOOL DYEING WITH ACID DYES

I.Yu. Vashurina, Yu.A. Kalinnikov

An original approach is proposed for the experimental investigation of interparticle interactions in the system consisting of dye solution and textile polymer during wool dyeing with acid dyes. This approach gives the possibility to interpret the mechanism of dye adsorption in aqueous-salt media. The role of ionic and non-ionic interparticle interactions has been differentiated and analyzed in conformity with the processes of dye solution formation and dye transfer from the solution into the polymer.

The authors have undertaken to interpret (experimentally) the mechanism of the intensification of acid dye adsorption by wool fiber in aqueous-salt media. For this purpose, a two-stage simulation of the whole dyeing process has been performed. The first stage was devoted to the investigation and elucidation of the mechanism of the interaction of real dyes with a wool keratin model. The second stage concerned model dye adsorption in the real fiber. Such model substances have been chosen that can take part only in interactions of a non-ionic type. This selection principle of the objects under investigation allowed the authors to reveal the role of non-ionic interparticle interactions both in dye solutions and in dye adsorption by wool from aqueous-salt media. The application of this principle also provided the opportunity to examine all physicochemical properties of the complex heterogenic dyeing system, using a general theoretical position based on modern ideas about aqueous solution structure. Until now, these ideas have been inadequately introduced into wool dyeing theory. However, by using this approach, the authors have succeeded in elucidating the contribution and mechanism of non-ionic interparticle interactions within the traditional dye adsorption theoretical framework where ionic ideas predominate.

O-xylene, which does not mix with water and slightly dissolves acid dyes, was chosen as a fiber model. The properties of this organic solvent, together with the solvent's molecular structure, give the rationale to consider o-xylene to be a wool keratin macromolecule elementary link model. The use of a solid model substrate in an experiment does not seem expedient because the polymers that do not possess ionic groups are rigid-chained and remain difficult to stain in dyeing conditions.

The procedure of the investigation of salt influence on the acid dye distribution between the solution and non-ionic wool polymer fragments was grounded on the analysis of dye distribution between the two conjugated liquid phases: aqueous salt solution and organic solvent. The essence of the experiment lay in estimating the hydrophilic-hydrophobic ratio that is the empirical parameter suggested in [1]. This parameter shows how much stronger is the inclination of dye to stay in water in comparison with its disposition to transfer into an organic medium where ionic interactions are impossible.

The amount (constant in all the experiments) of purified acid dye (C.I. Acid Orange 7) deliberately surplus for the dissolution was thoroughly mixed with a small quantity of water (containing salt addition) in a glass vessel. Then the whole system was thermostated at 293 K for twenty-four hours. As a result, a saturated dye solution over a thick sediment layer was obtained. Then an equal volume of o-xylene was added to this solution. The vessel was shaken for a long time and then kept at the same temperature up to the achievement of an equilibrium state (twenty-four hours more). When all of these operations had been completed, probes of the aqueous and organic phases were taken, and the dye content was defined spectrophotometrically. Three chlorides of multi-valent metals (aluminum, zinc, and cadmium) were included into the range of additions under investigation. Such a choice was conditioned upon the following circumstances:

- the extremely high efficiency of $AlCl_3{\cdot}6H_2O$, $ZnCl_2$, and $CdCl_2{\cdot}2.5\ H_2O$ as the reagents, which increase acid dye equilibrium adsorption by wool;

- the significantly expressed specificity of these particular salts in their concentration influence on dye yield in the fiber [2];

- the existence of the results of mathematical analysis of the cited salt-containing dyeing systems from a position of ion exchange [3].

The results of the research are presented in Fig. 1-3. As Fig. 1 demonstrates, when the salt concentration in water is increased, the solubility of the acid dye diminishes consistently with the diminution being especially sharp (30-55%) in the salt

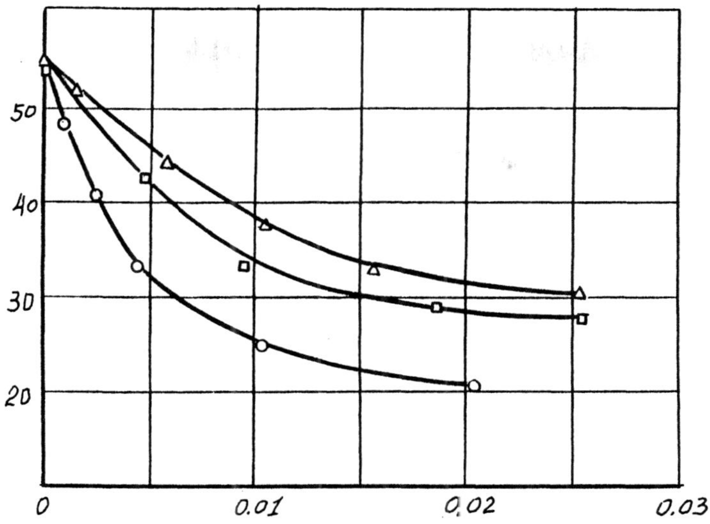

Fig. 1. The solubility of C.I. Acid Orange 7 in aqueous solutions of cadmium (1), zinc (2), and aluminum (3) chlorides at 293 K.
X = Salt concentration (c_{salt}), mol/l; Y = Solubility of dye ($L_{aq.}$), g/l.

concentration interval 0 - 0.01 mol/l. The picture thus obtained can be easily explained on the basis of traditional notions of the electrolyte influence on the state of ionic dyes in solution. The electrolyte action is displayed in the formation of ionic pairs between dye anion and electrolyte cation as well as in the generation of dye associates (These can include either only organic anions or, additionally, counter-ions, i.e. metal cations [4-6]). The higher the charge of salt cation, the more intensive is the stimulation of these interactions by the electrolytes.

While the result of salt influence on acid dye solubility in water is logical and predictable, the consequence of salt action on dye transfer into the organic liquid phase is perfectly amazing (Fig. 2). All the curves reflecting the variations in acid dye content in o-xylene as a function of salt concentration in the conjugated aqueous phase are characterized by the existence of keen maxima. For a water-aluminum chloride system, this maximum corresponds to a salt concentration of 0.0005-0.0007 mol/l and for zinc and cadmium chloride solutions 0.004-0.007 mol/l. In the process of analyzing the reasons for the harsh increase (2.1-2.5 times) of acid dye solubility in o-xylene (begotten

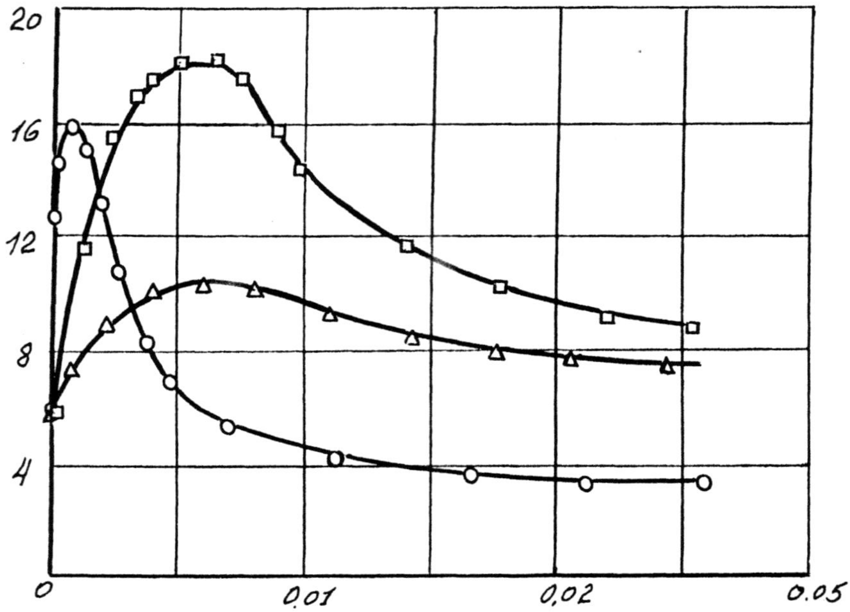

Fig. 2. The influence of cadmium (1), zinc (2), and aluminum (3), chloride concentrations in the saturated aqueous solutions of C.I. Acid Orange 7 upon the transfer of dye into the organic solvent.
X = Salt concentration (c_{salt}), mol/l; Y = Solubility of dye ($L_{org..}$), g/l.

by salt introduction into the conjugated aqueous phase), one must start from the fact that the powder form of the acid dye is insoluble in benzene and its derivatives. Nevertheless, the concentration of Acid Orange 7 in o-xylene reaches 6 g/l when the dye transforms into the organic liquid from water (i.e. having been hydrated) and increases up to 11-17 g/l from water-salt systems. Consequently, maximal dye content in the organic solvent is defined by the state of the dye in the aqueous medium from which the diffusion takes place.

The addition of inorganic electrolytes into ionic dye aqueous solutions, as has been already pointed out, inducts the interactions dye - dye and dye - salt. Ionic adducts formed by dye anions and metal cations can be either charged (when positive and negative charges are not compensated (a)), or uncharged (b):

$[Dye\text{-}SO_3^- \ M^{2+}]^+$ (a)

$[Dye\text{-}SO_3^- \ M^{2+} \ {}_3^-OS\text{-}Dye]$ (b)

The latter can be considered as pseudo-non-electrolytes. Therefore, their stability in water would be much lower (and vice versa in organic liquids, much higher) than the solubility of dyes in ionic form. Obviously, the bigger the proportion of uncharged particles in the general balance of all dye forms in its saturated aqueous solutions, the greater is that part of the dye that transforms into the organic phase.

During the course of the gradual enhancement of salt concentration in an aqueous composition, oppositely charged ions attract each other first of all [6,7]. Then the formation of dye associates (firstly, small ones, later on, larger and larger ones) takes place. Therefore, just small additions of salts stimulate acid dye transfer from water into the organic phase while a further increase of electrolyte content in the solution gradually levels the achieved effect. From the kinetic perspective, the transfer of a part of dye (in the form of uncharged ionic adducts) into the organic solvent (when it comes into contact with water) causes the decrease of this form concentration in an aqueous solution. This decrease leads to the disintegration of large associates and to the formation of new pseudo-non-electrolyte units. The latter again partly transfer into the organic phase with this process continuing up to the establishment of an equilibrium dye distribution between water and the organic liquid. As Fig. 2 demonstrates, the introduction of zinc chloride into the aqueous dye solution provides the maximal solubility of Acid Orange 7 in the conjugated o-xylene phase. This described regularity is appropriate because the particle consisting of a Zn^{2+}- ion and two dye anions (in comparison with the huge uncharged particle, including the Al^{3+}- ion and three dye anions) possesses comparatively low mass and is capable of existing in the organic solvent. The dye adduct with a Cd^{2+}- ion is significantly larger than the dye adduct with a Zn^{2+}- ion (due to the huge dimensions and mass of cadmium) and, consequently, less soluble.

Thus, the first stage of the present work has proved experimentally that hydrophilic-hydrophobic acid dye properties are stipulated not only by dye molecule structure [1, 8-16] but also by the electrolyte composition of dyeing solutions. This fact is illustrated by the curves in Fig. 3 that reflect the influence of salt concentration on the ratio of dye

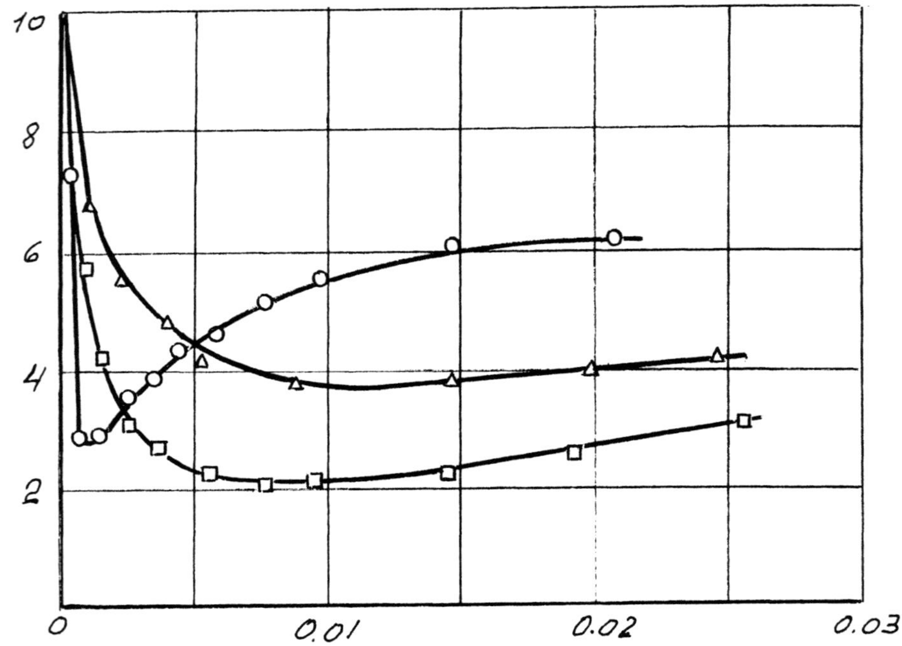

Fig. 3. The influence of aluminum (1), cadmium (2), and zinc (3) chloride concentrations upon the ratio of C.I. Acid Orange 7 solubilities in two conjugated liquid phases (water-salt and organic).
X = Salt concentration (c_{salt}), mol/l; Y = $L_{aq.}/L_{org}$

solubilities in the conjugated aqueous and organic media. The region of low salt contents (up to 0.0007 mol/l in the case of an $AlCl_3 \cdot 6H_2O$ application and up to 0.007 mol/l in the case of $ZnCl_2$ and $CdCl_2 \cdot 2.5H_2O$) is marked by an abrupt drop of dye hydrophility. Further growth of the concentrations of the investigated salts changes the tendency of a hydrophilic-hydrophobic ratio increase into the opposite, with aluminum chloride giving the most expressive influence.

If one compares the results discussed in this article with the ones obtained earlier in the course of the evaluation of the equilibrium adsorption of Acid Orange 7 by wool from water-salt systems [2] (for convenience's sake, the results in Fig. 4 are represented in the percents of the initial dye content in the residual bath), the analogy between the salt influence upon dye transfer both into the solid substrate phase and into the organic

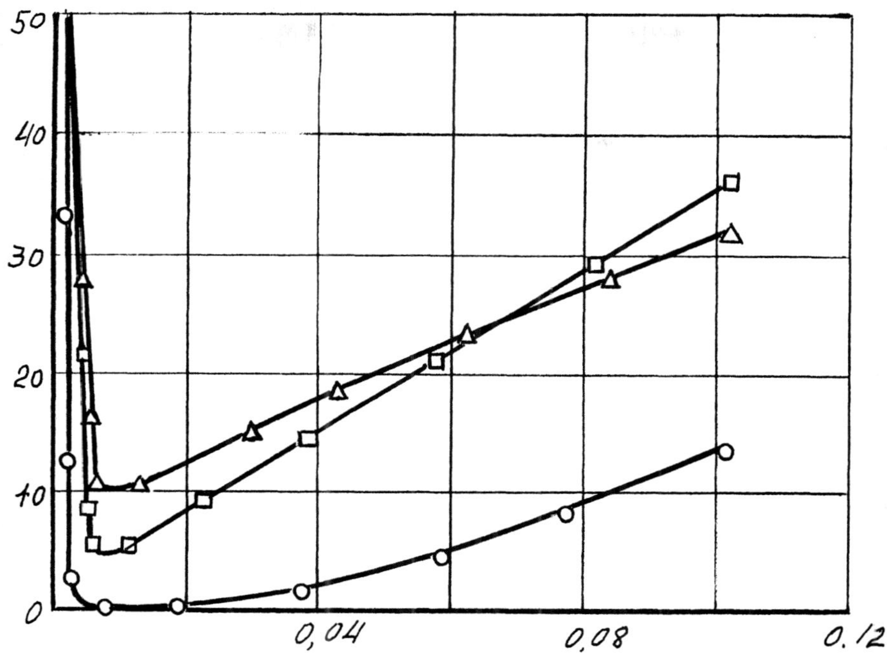

Fig.4. The influence of zinc (1), cadmium (2), and aluminum (3) chloride concentrations in aqueous dye solutions upon C.I. Acid Orange 7 equilibrium adsorption by wool fiber.
X = Salt concentration (c_{salt}), mol/l; Y = Dye content in the residual bath, %.

solvent phase becomes obvious. This similarity allows us to confirm that the intensifying influence of multi-valent metal chlorides on acid dye adsorption in wool fiber is substantially conditioned by the stimulation of non-ionic interactions between the dye and the polymer.

At the second stage, the simulation of acid dye equilibrium adsorption in wool has been undertaken by applying non-ionic acid dye models. The best models of acid dyes are disperse dyes because they are characterized by a similar molecular structure but do not possess ionic properties. Three disperse dyes were used in the experiment: one of them was an azo-compound (C.I. Disperse Orange 31) while the other two were antraquinone derivatives (C.I. Disperse Violet 1 and C.I. Disperse Red 15).

The dyeing of wool fiber patterns was performed at boiling temperature in round-bottomed flasks fitted with reflux condensers with the solution ratio being 1 : 50. The disperse dye content in water was constant and equal to 0.04 g/l (This corresponds to 2%

of dye mass relative to fiber mass). Salt concentrations were varied from 0 to 0.02 mol/l. The pH -index was stabilized at 3.2 using small additions of highly concentrated HCl. To estimate the quantity of dye adsorbed by the fiber, the residual bath was evaporated dry in mild conditions (firstly, using a water bath and thereafter by applying hot air at 65-70°C). Then the dye obtained was dissolved in 10 ml of acetone, and the solution was analyzed colorimetrically. Similarly, the dye content in the rinsing waters was defined. The scouring of every fiber pattern in water was carried out (like dyeing) by means of a prolonged boiling of the solution, with its volume being maintained constant.

The results of the described experiment are shown in Fig. 5-7. The curves on the main graph characterize dye content in the scoured wool fiber. Although in this investigation, three salts were used (as in the first stage), $AlCl_3{\cdot}6H_2O$, $ZnCl_2$ and $CdCl_2{\cdot}2.5H_2O$, the dependencies are given which describe dye behavior only in the presence of the first two chlorides. The exclusion of cadmium chloride from this discussion is due to wool color change while dyeing from the solution containing this particular salt.

The common character of the relations reflecting salt influence on the effectivity of dye output is clearly recognized (the system Disperse Red 15 - aluminum chloride being an exception). Small additions of aluminum and zinc chlorides (0.0004-0.0007 mol/l for Disperse Orange 7 and 0.001-0.005 mol/l for Disperse Violet 1) cause substantial (for 20-200%) growth of the amount of dyes fixed in the fiber. It is noticeable that the regularities of dye fixation degree alterations are identical in shape to those just pointed out. A subsequent increase of salt content in the dyeing solutions leads to the reduction of disperse dye adsorption down to the initial values (or even smaller). However, when salt concentration is further increased, the rise of dye adsorption starts again, at this time now being consistent. It must be stressed once more that the shape of dye adsorption isotherms (at a constant dye concentration and with a variable salt content) is similar for adsorbed dye (before wool scouring) and for fixed dye (after wool scouring). This experimental fact suggests the idea that water-salt systems, which provide greater dye output in the fiber, ensure a different (in comparison with pure aqueous dye solutions) quality of adsorbed dye. In our opinion, the increase in disperse dye solubility under the influence of the investigated salts is accompanied by the accumulation (in aqueous

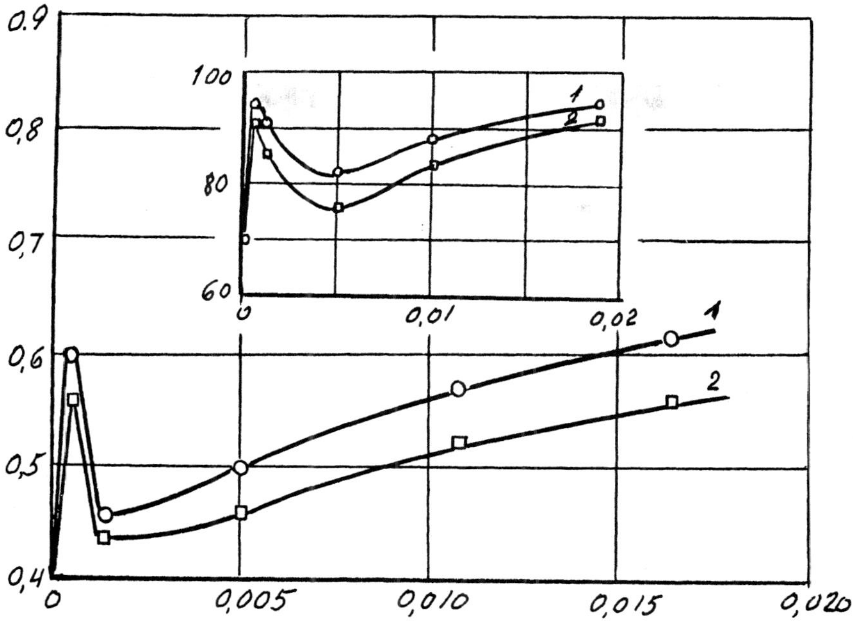

Fig. 5. The influence of aluminum (1) and zinc (2) chlorides in the dyeing solutions on the equilibrium adsorption and fixation degree of C.I. Disperse Orange 7 in wool fiber.
X (x) = Salt concentration (c_{salt}), mol/l; Y = Dye content in wool fiber, g/kg; y = Dye fixation degree, %.

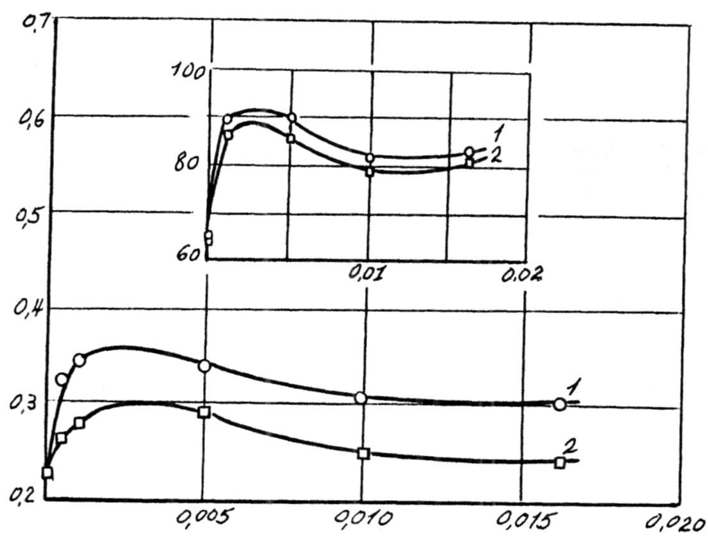

Fig. 6. The influence of aluminum (1) and zinc (2) chlorides in the dyeing solutions on the equilibrium adsorption and fixation degree of C.I. Disperse Violet 1 in wool fiber.
X (x) = Salt concentration (c_{salt}), mol/l;
Y = Dye content in wool fiber, g/kg;
y = Dye fixation degree, %.

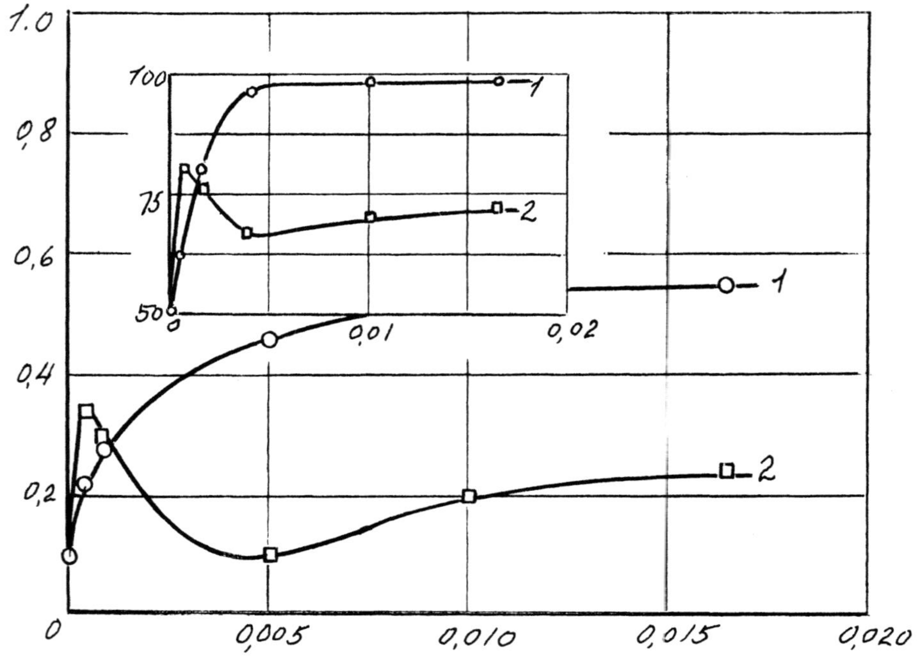

Fig. 7. The influence of aluminum (1) and zinc (2) chlorides in the dyeing solutions on the equilibrium adsorption and fixation degree of C.I. Disperse Red 15 in wool fiber.
X (x) = Salt concentration (c_{salt}), mol/l; Y = Dye content in wool fiber, g/kg; y = Dye fixation degree, %.

solution) of such dye form which possesses a heightened affinity to wool and therefore is selectively adsorbed and durably fixed in the fibrous polymer.

In conformity with modern ideas, dye transfer into the fiber (irrespective of the nature of the fiber) proceeds by the adsorption of the dissolved dye fraction [17]. That is why to understand the causes of the stimulating effect produced by some inorganic salts on dye adsorption, the influence of these salts on the solubility of disperse dyes in water was studied.

Small amounts (0.1 g/l) of purified disperse dyes were placed in hermetic glass vessels and filled with aqueous salt solutions with various concentrations of aluminum, zinc, and cadmium chlorides. The contents were stirred and thermostated at 293 K for two days until an equilibrium state was achieved. Dye solubility was estimated according to the change in the optical density of dye solutions as the salt concentration in them was raised. A part of the data obtained (particularly those concerning aluminum-chloride-holding dye solutions) is presented in Fig. 8. Firstly, the striking influence of aluminum chloride on the solubility of all three disperse dyes is evident. The characteristic maximum in dye content at small concentrations of salts (containing cations structure-breakers) is similar to those obtained for more simple organic non-electrolytes than dyes [18]. This maximum corresponds to the region of the greatest disorder in water structure and is conditioned upon the strengthening of universal and specific dye molecule solvation [19, 20]. An analogous maximum is also present on the curves characterizing the solubility of Disperse Orange 7 and Disperse Violet 1.

The second rise in the solubility of the latter two dyes (during the increase of the aluminum chloride concentration in the solution) can be connected with the interaction between the metal cations and the polar functional groups of disperse dye molecules [21]. This interaction results in the increase of dye molecule activity in the adsorption process (in conformity with the mechanism proposed and described in [22, 23]). According to this mechanism, the higher the dipole moment of dye molecule, the stronger is the intensifying influence of cations on dye adsorption. That is why the change in the solubility of Disperse Orange 7, whose azo-chromophoric body is lengthy and highly polar, is so large. Conversely, this change is only slightly outlined for Disperse Violet 1, whose antraquinone derivative molecular body is compact and low polar. The curves, describing the dependencies of the optical density of disperse dye saturated solutions upon the content of $ZnCl_2$ and $CdCl_2{\cdot}2.5H_2O$ in aqueous media, are completely analogous to those with $AlCl_3{\cdot}6H_2O$, but in the graphs they are located lower due to the less structure-breaking effect of zinc and cadmium in comparison with aluminum. The relation, reflecting the influence of $AlCl_3{\cdot}6H_2O$ upon the optical density of Disperse Red 15 aqueous solutions (Fig. 8, curve 1), differs radically from all relations previously described. In the concentration interval studied, the dye solubility increase is directly proportional to the salt content. As was shown in [19], this regularity is connected with the formation of a complex between three molecules of Disperse Red 15 and one Al^{3+} - cation.

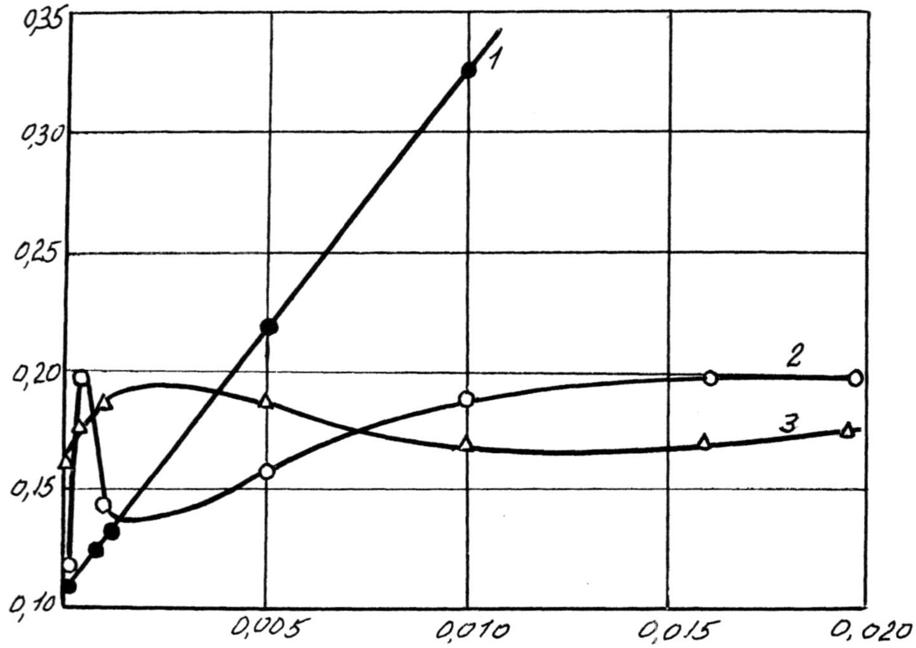

Fig. 8. Aluminum chloride influence on the optical density of saturated aqueous solutions of the following dyes: C.I. Disperse Red 15 (1), C.I. Disperse Orange 7 (2), C.I. Disperse Violet 1 (3)
X = Salt concentration (c_{salt}), mol/l; Y = Light absorption of dye solution (optical density).

The revealed diversity of aluminum chloride influence upon the solubility of disperse dyes (depending on the peculiarities of dye molecule structure and salt concentration in the solution) (Fig. 8) rightly corresponds to the alteration of the shape of dye adsorption isotherms (Fig. 5-7). As a consequence, the dissolved dye fraction determines the efficiency of dye adsorption by wool: not only in the case where a dye molecule is solvated according to the known mechanism [17], but also in the case where dye solubility in water is conditioned by other reasons such as the interaction of dye polar functional groups with inorganic electrolytes or even a complex formation.

The absence of the proportional interrelation between dye adsorption increase and the augmentation of dye solubility in water-salt systems (Fig. 9) gives an indication of the existence of definite restrictions in fiber adsorption capacity. This fact is confirmed by the Acid Red 15 adsorption isotherm which demonstrates the achievement of maximal equilibrium color yield when the concentration of $AlCl_3 \cdot 6H_2O$ equals 0.01 mol/l, whereas the dye solubility continues to rise even at higher salt concentrations. The maximal possible output of Disperse Red 15 (corresponding to the horizontal region of the isotherm 2 in Fig. 9) is approximately 0.0021 g-equiv/kg. At the same time, while not being the highest (see curve 3 in Fig. 9), the Disperse Orange 7 yield in experimental conditions reaches 0.0026 g-equiv./kg. Such a regularity derives from the fact that the dye C.I. Disperse Red 15 exists in the solution (and hence is adsorbed by the fiber) in the form of a complex that includes three dye molecules. It is very difficult for such a huge particle to be fixed in the fiber because of steric restraints, whereas it is known that all the molecular (and also hydrophobic) forces can only be displayed when dye molecules directly adjoin non-ionic fiber fragments. The above-said obstacles are absent in the case of the adsorption of Disperse Orange 7 that is either slightly associated or even not associated in aqueous salt solution.

Based on the elucidated regularities of the influence of multi-valent metal chlorides upon disperse dye behavior in aqueous solutions and upon the dye adsorption process, one can make a proposal about the mechanism of the intensification of dye stuff transfer from the solution into the textile polymer in the presence of small additions of the afore-cited salts. We can guess that the general cause of the observed phenomena is concealed in the change of the disperse dye solutions that is begotten by breaking the aqueous solution structure due to the introduction of small amounts of multi-valent cations. In more detail: those organic molecule parts which in the absence of salts are hydrated hydrophobically (with water clusters participating in the solvation), in the presence of salts (which destruct water clusters and polarize water molecules) become hydrated by separate water molecules. This re-solvation results in the rise of dye solvation as a whole and consequently in the heightening of dye solubility. But the dye molecule stops being "fitted" into the water structure. The solvation shell becomes more mobile and easily abandons the dye molecule when the dye molecule comes in contact with the polymer. Thus, the more destructed the water structure, the more effective is the interaction between the adsorbing dye and the fiber (It must be taken into consideration that the above-said salt influence applies also to non-ionic wool fragments.).

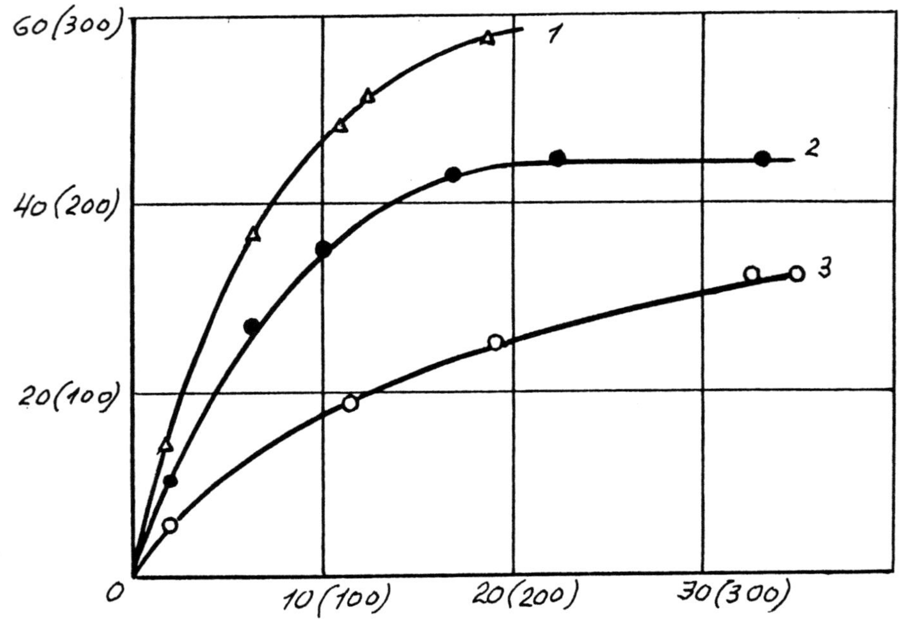

Fig. 9. The correlation between the increase in optical density of disperse dye saturated aqueous solutions (under the influence of aluminum chloride) and the augmentation in dye equilibrium adsorption by wool from the corresponding water-salt media:
1 - Disperse Violet 1;
2 - Disperse Red 15 (The designations on the axes in brackets);
3 - Disperse Orange 7.
X = The increase in optical density of saturated dye solutions, %;
Y = The increase in dye content in the fiber, %.

As appears from the above, both stages of this research demonstrated the effect of the stimulating influence of salts on the interactions between the dyes and the fiber. The cause of this phenomena lies in the change of non-ionic fragment solvation of acid dyes and wool fiber due to water structure break produced by multi-valent metal cations.

References

1. J.R. Haugen and E.R. Hardwick, J. Phys. Chem, 67 (1963) 725.
2. I.Yu. Vashurina et al., Izv. Vuzov. Technol. Tekstil. Prom., (1) (1994) 44 (in Russian).
3. F.Yu. Telegin et al., J. Phys. Chem., 67 (4) (1993) 752 (in Russian).
4. T. Vickerstaff, The Physical Chemistry of Dyeing (Moscow: Gizlegprom, 1956) (Russian Edition).

5. R.H. Peters, Textile Chemistry. The Physical Chemistry of Dyeing (Amsterdam, Oxford, N.-Y.: Elsevier Sci. Publ. Co., 1975).
6. E. Coates, J. Soc. Dyers and Col., 85 (8) (1969) 355.
7. M. De-Vijlder, J. Chem. Faraday Trans., 79 (1) (1983) 155.
8. J.B. Speakman and G.H. Elliott, J. Soc. Dyers and Col., (59) (1949) 185.
9. W.G. Grewther and L.M. Dowling, J. Text. Inst., (51) (1960) 775.
10. J.M. Gillespie et al., J. Text. Inst., (5) (1960) 703.
11. J. Steinhardt, C.H. Fugitt and M. Harris, J. Res. Nat. Bur. Standarts, (26) (1941) 293.
12. Pfeiffer, Organische Moleculverbindungen (Stuttgart, 1932).
13. G.A. Gilbert, Prac. Roy. Soc. Ser. A, (183) (1944) 167.
14. H. Zollinger, J. Soc. Dyers and Col., 81 (8) (1965) 345.
15. I. Klotz and J. Urquhart, J. Amer. Chem. Soc., (71) (1949) 847.
16. L. Peters and G.H. Lister, Faraday Discussion on Dyeing and Tanning, (16) (1959) 24.
17. G.A. Krestov, The Thermodynamics of Ionic Processes (Leningrad: Khimiya, 1984) (in Russian).
18. S.M. Prusova, The State of Disperse Dyes in the Solution and the Ways for the Intensification of the Processes of Polyester Dyeing (Dissertation for the degree of Candidate of technical science, Ivanovo, 1990) (in Russian).
19.A.M. Pron'kin, Yu.A. Kalinnikov, B.N. Mel'nikov, J. Appl. Chem., 62 (5) (1986) 1116 (in Russian).
20. H. Peckarski and M. Tkaczuk, J. Chem. Soc. Faraday Trans., (22) (1991) 3661.
21. I.Yu. Vashurina, Yu.A. Kalinnikov and I.A.Prokopenko, Izv. Vuzov. Tekhnol. Tekst. Prom., (1) (1993) 53 (in Russian).
22. Yu.A. Kalinnikov , The Intensification of Textile Material Dyeing by Means of the Influence on the Solvation and Association Processes in Dye Solutions and the Creation of Highly Effective, Environmentally Safe Dyeing Methods (Dissertation for the degree of Doctor of technical science, Ivanovo, 1993) (in Russian).
23. I. Sivaraja, Colourage, (17) (1987) 15.

The Institute of Chemistry of Non-Aqueous Solutions
of the Russian Academy of Sciences, Ivanovo

THE FERMENTATION DESTRUCTION OF LIGNIN IN COTTON AND FLAX

A.V.Cheshkova, V.I.Lebedeva, B.N.Mel'nikov

The chemism of fermentative destruction of lignin-containing admixtures in cotton and flax has been studied using UV- and IR-spectroscopy methods. It has been found that white rot fungi containing oxidoreductases cause the removal of 16-66% of native lignin from cotton husk and from flax. The residual lignin is characterized by substantial changes in chromophoric and aromatic structures.

Among the concomitant substances of native cellulose, lignin is the one of the most difficult to remove. To a large extent, the degree of lignin component removal defines the conditions of the processes of textile material finishing. Lignin is resistant to the influence, not only of chemical reagents, but also of microorganisms. Nevertheless, it is known that some ferments possess lignin destructing capacity. Basidium white rot fungi which synthesize oxidoreductases [1, 2] belong to these microorganisms. With the development of bio-technologies, basidiomycetes will become the most accessible source of ferments of different types on an industrial scale. However, the introduction of bio-processes into textile processes requires thorough research.

The aim of the present work consisted in researching the bio-destruction of lignin-containing admixtures of cotton and flax under the influence of basidiomycetes grown using available nutrient media including industrial wastes. The following strains-producers were chosen for our investigation: *Cerrena maxima Murr* (which synthesize peroxidase), *Coriolius hirsutus* (containing presumably peroxidase), and *Coriolius Poliporizae* (complex preparation of peroxidase and laccase). To study the degree and depth of lignin fermentative destruction, UV-spectroscopy was used. The residual lignin content in the substrates investigated was evaluated according to the UV-spectra of lignin dioxane extracts. The structural changes in lignin were estimated using IR-spectroscopy [3, 4].

Real biological substrates were used in the experiment: cotton fabric (calico, article 46), linen fabric, cotton husk, isolated sulfuric lignin, and the lignin model (o-vanilline).

The treatment of the enumerated substrates with the solutions of ferments (concentration 1 g/l) was performed at a constant temperature of 50^0C with a pH of 7.5 for 30 minutes. In addition to the ferments, these solutions contained a non-ionic wetting agent with a concentration of 0.1 g/l and a stabilizer.

The results obtained indicate the high activity of the chosen ferments in lignin decomposing processes. Thus, the use of peroxidase enables the removal of up to 44.9% of the lignin component from the fabric; the use of polyphenoloxidase of up to 41.3% of lignin; and the use of oxidoreductase of approximately 16.0%. Bio-degradation of the lignin component increases when the ferments of polyphenoloxidase and peroxidase are introduced into the solution simultaneously. The degree of lignin removal thus achieved equals 66%. This figure is greater than that obtained after fabric alkaline scouring (44%). What's more, the treatment with ferments eliminates the destruction of cotton cellulose because the influence of ferments is highly selective [2].

Fig.1 demonstrates the results of the fermentative treatment of cotton husk (Fig.1-a) and of flax sulfuric lignin (Fig.1-b). In both cases, the maximum at 280 nm (conditioned by light absorption by aromatic lignin structure) can be easily seen. This maximum corresponds to the allowed π-electron transition of benzoyd forms. The electron transfer

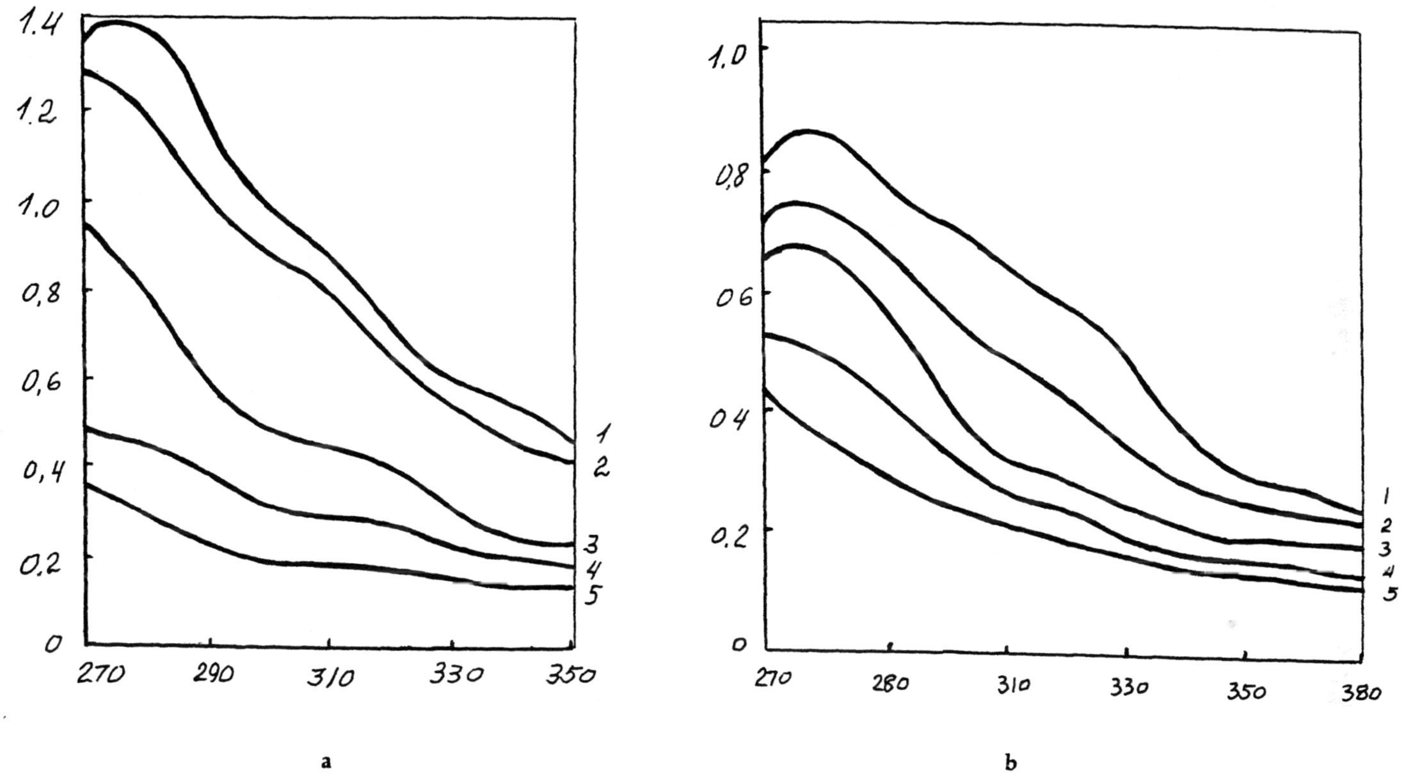

Fig.1. UV-spectra of dioxane extracts of lignin from flax (a) and cotton husk (b):1 - original lignin; 2, 3, 4 - lignin treated with oxidoreductase, polyphenoloxidase,and peroxidase, respectively; 5 - lignin treated with the composition of oxireductases.

X = λ, nm;

Y = Light absorption (optical density of the solution).

of a $n \rightarrow \pi^*$ type (which results in the formation of a "shoulder" on the spectral curve at 360-370 nm) corresponds to the lowest energy of singlet - singlet molecule excitation.

As is seen from the spectra presented, the ferments under investigation cause a hypochromic displacement of lignin short-wave maximum. This displacement is accompanied by a general decrease in light absorption within the long-wave spectral region. The last of the described regularities is indicative of the destruction of the lignin macromolecule conjugative system and of sizable changes in lignin chromophoric structure.

Lignin destruction processes, as has been already mentioned, are more effective when redox ferments for the lignin treatment are used in conjunction with one another. On one hand, this is connected with the existence of a synergetic effect in the processes where different ferments react simultaneously. On the other hand, this reveals the distinction between the mechanisms of lignin decomposition by different ferments (Fig.1-a, b, curves 5), and this distinction must originate in the structural peculiarities of the lignin macromolecule itself.

UV-spectra of o-vanilline which can be considered as the simplest model of carbonyl-containing structural links of lignin macromolecules are presented in Fig.2 and 3. The treatment conditions of o-vanilline were similar to those described for lignin substrate. The results show that the reaction of o-vanilline destruction is almost completed just at the beginning of the bio-oxidizing process (Fig.3). Already within the first three minutes of treatment, an abrupt drop of light absorption in the spectral regions at 270-280 nm and 360-380 nm is observed. The difference between the magnitude of the changes in light absorption at short-wave and long-wave maxima gives proof of the less intensive destruction of benzene structures in comparison with functional groups during vanilline

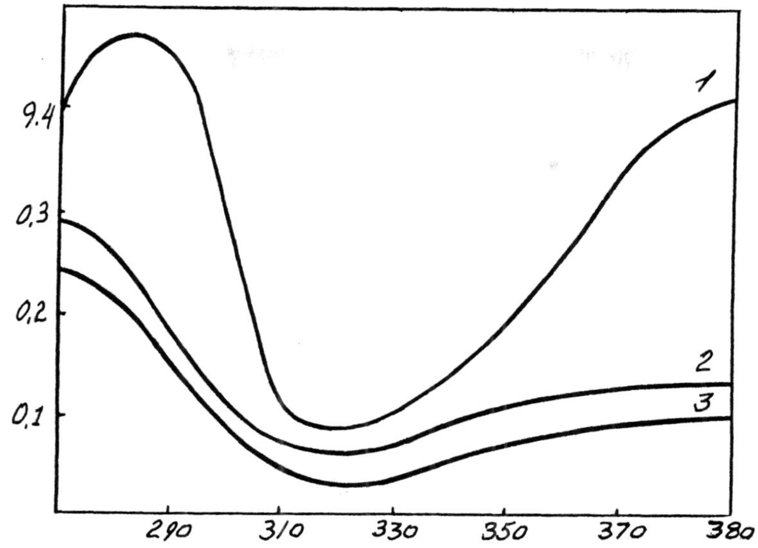

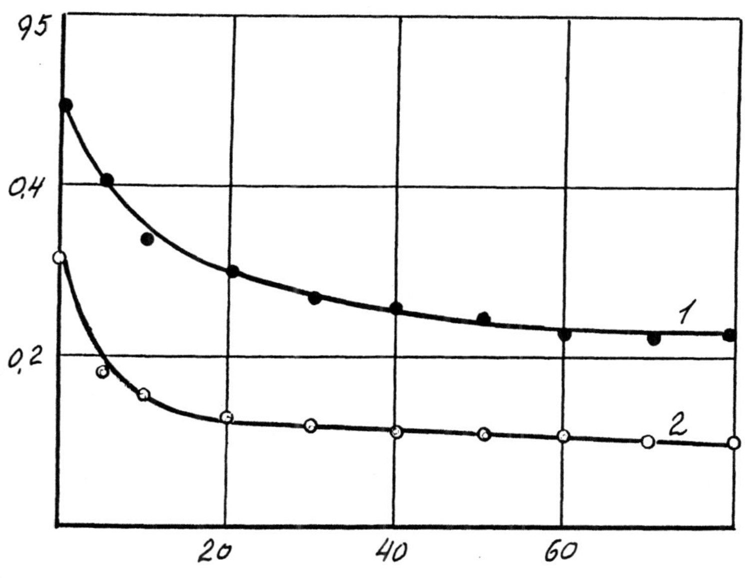

Fig.2. UV-spectra of the original o-vanilline (1) and of o-vanilline treated with oxidoreductase (2) or peroxidase (3).

X = λ, nm;

Y = Light absorption (optical density of the solution).

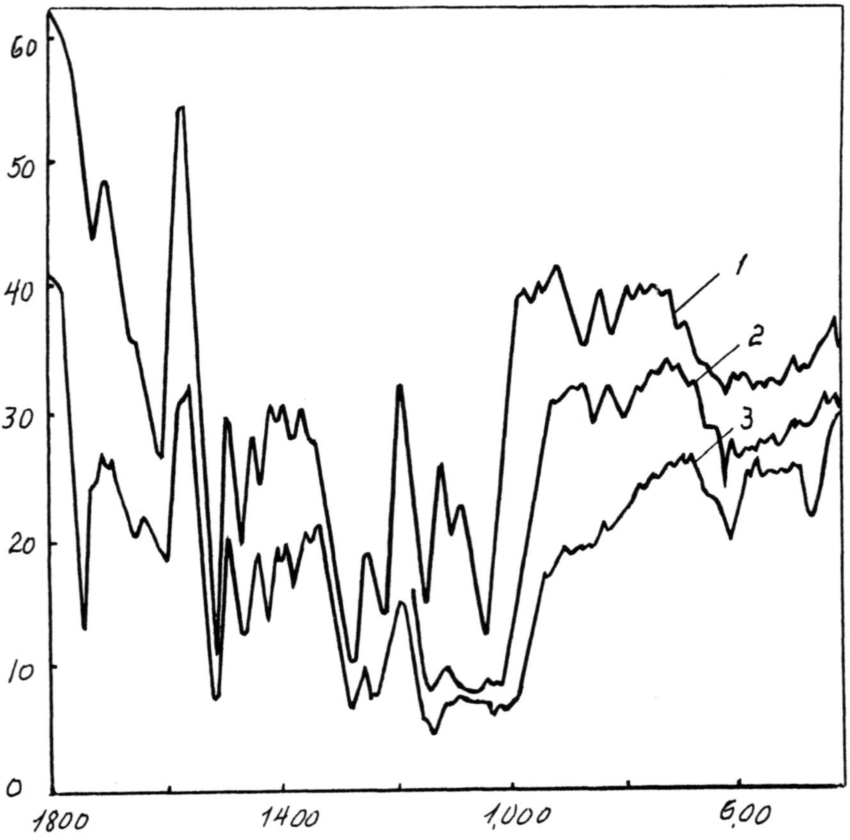

Fig.3. The alterations of the intensity of light absorption by o-vanilline at 280 nm (1) and at nm (2) in the course of its treatment with peroxidase.

X = time, min.;
Y = Light absorption (optical density)

bio-treatment. The existence of a hyperchromic effect at 280 nm is obviously connected with the following reversible deoxidization reaction:

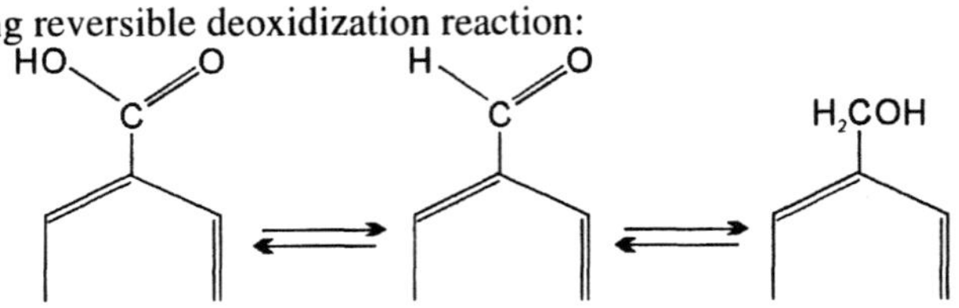

The considerable decrease in light absorption at the long-wave part of the spectra in Fig. 2 (and almost complete disappearance of long-wave maxima after treating with peroxidase) is indicative of the destruction of the o-vanilline chromophoric system. This fact gives the possibility to suppose that the ferments under discussion cause the intensive process of o-vanilline demethylating:

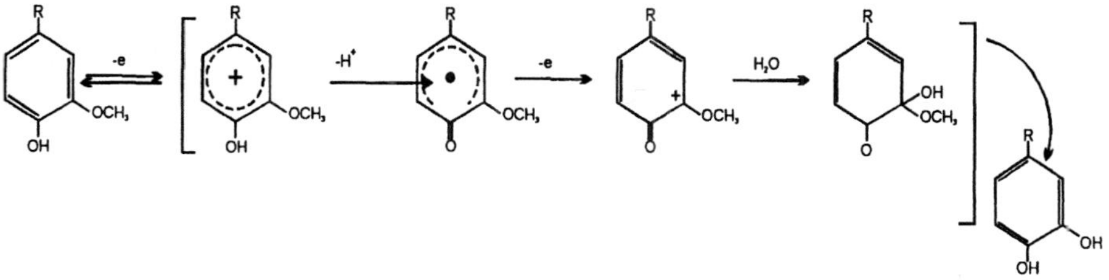

The correlation of the results obtained by UV- and IR-spectroscopy methods is observed. A comparison of IR-spectra of the original lignin and residual lignin preparations shows that exoenzymes of the strains under investigation provoke sizable changes in the functional composition of lignin and in lignin polymeric structure. The decrease in methoxy-group content after lignin is treated with peroxidase is noticed. Oxidoreductase causes the complete decomposition of lignin and this is confirmed by the abrupt lowering of the absorption at 1088 cm^{-1}. On the spectra of residual lignin, a new band at 1720 cm^{-1} appears which is conditioned by the formation of carboxyl-groups.

The alteration of the intensity of light absorption at the region of 1000-1300 cm^{-1} attracts attention. The spectrum of original lignin in the mentioned interval is characterized by the presence of four bands at 1040, 1120, 1220, and 1290 cm^{-1}, stipulated by the oscillation of ester and ether bonds. On the spectral curves of residual lignin, a harsh decrease in the intensity of characteristic frequencies under consideration takes place caused by the break up of lignin ether bonds while carbonyl groups remain untouched. Thus, lignin decomposition includes the reactions of demethylating and of breaking the less stable β-ether bonds. In these reactions, a ferment plays the role of electron donor and of hydrogen carrier from donor to acceptor. The destruction of lignin by the investigated redox ferments can be presented by the following conventional pattern:

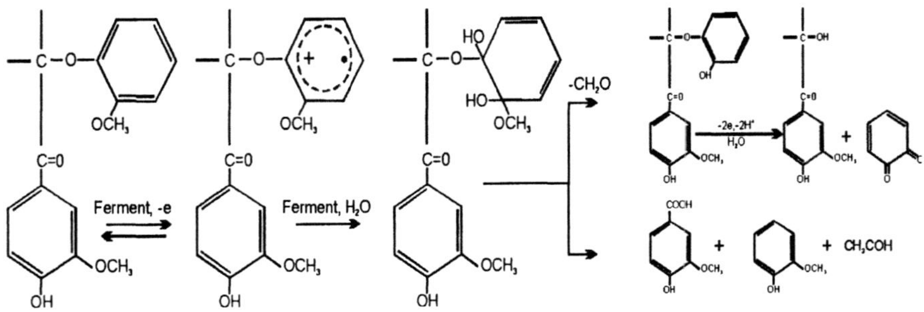

The disappearance of the band at 1590 cm^{-1} in the residual lignin IR-spectra and the almost complete straightening of UV-spectral curve, which characterizes the results of the simultaneous treatment of lignin with two ferments (Fig.1-a, b, curves 5), give proof of the capacity of the investigated ferments to destruct stable carbon-carbon bonds in lignin macromolecules. Obviously, the bio-degradation of lignin is accompanied by the rupture of $C\alpha - C\beta$-bonds in the propane chain.

The catalytic activity of basidiomycetes is basically connected with the presence of imidazole-groups [5]. The reaction of the bio-degradation of structural links goes through the stages of the formation of the ferment-substrate complex and the subsequent restoration of the ferment structure according to the following pattern:

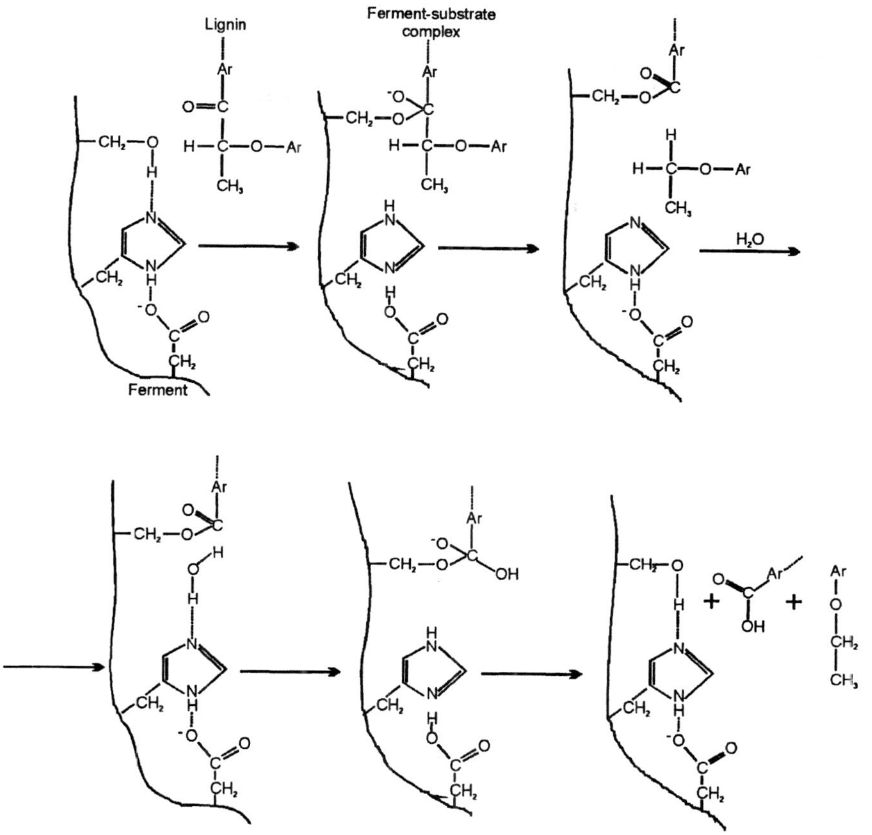

The changes in IR-spectra of residual lignin in the interval of 400-900 cm^{-1} are also noticed, and this can be attributed to the generation of low-molecular products of lignin destruction.

Conclusions

1. The process of fermentative destruction of cotton lignin and flax lignin has been studied using UV- and IR-spectroscopy.

2. It has been shown that basidium white rot fungi which contain oxidoreductases cause lignin bio-degradation and therefore ensure the removal of 16-44.9% of lignin from the cotton husk as a result of the husk's bio-treatment.

3. The mechanism of lignin bio-destruction by means of ferments has been proposed.

References

1. V.N. Kreinberg, N.R. Ozolinya and V.N. Sergeeva, *Khimiya Drevesiny,* (3) (1981) 69 (in Russian).

2. V.I. Lebedeva et al., *Izv. Vuzov. Tekhnol. Tekst. Prom.,* (5) (1989) 65 (in Russian).

3. V.M. Nikitin and V.P. Shchyogolev, in *Khimiya Drevesiny i Tsellyulosy* (Moscow: Lesnaya Promyshlennost, 1978) 151-153 (in Russian).

4. E. Nakanishi, in *Infra-Red Spectra and Structure of the Organic Compounds* (Moscow: Mir, 1965) 27 (Russian Edition).

5. Yu.E. Ovchinnikov, *Bioorganic Chemistry* (Moscow: Prosveshcheniye, 1987) (in Russian).

The State Academy of Chemistry and Technology, Ivanovo

A NEW UNDERSTANDING OF THE PROCESS OCCURING IN COTTON FIBER DURING TREATMENT WITH ALKAI-PEROXYDE SOLUTIONS

V.G. Stokozenko, V.N. Galashina, E.L. Kadykova, S.M. Gubina

The correlation between the content of functional groups in the cellulose material and the amount of aldehyde- and carboxyl-containing admixtures has been established. A new interaction mechanism between hydrogen peroxide and cellulose has been proposed with the specificity of the chemical structure and the energetic state of the anhydroglucose link taken into account. This mechanism explains the reasons for the non-correspondence between the decrease in the experimental value of cellulose polymerization degree and the stability of cellulose functional group composition.

In conformity with the classical theory of cellulose oxidizing destruction, the oxidizing of carbonyl groups (belonging to the second, third, and sixth carbon atoms) up to the carbonyl ones is the primary act of cellulose interaction with hydrogen peroxide in an alkaline medium. The accumulation of these groups is accompanied by a weakening of the β-glycoside links and their rupture due to alcohol dehydration and aldole condensation [1-3].

Beginning from this point, it is reasonable to presuppose the correlation between the enhancement of the oxidized cellulose functional group content and the lowering of polymerization degree (PD). But an analysis of the extensive scientific literature shows that the data concerning the functional group content in oxidized cellulose are extremely discrepant, and the dependence between the changes in group content and PD has not been generally noticed.

In the research, undertaken by Kaverzneva [2], it was established that an increase in carboxyl group content of up to 0.30-0.34% causes a cellulose PD decrease up to 200-

300. In [4], it is mentioned that after cellulose oxidizing by the same composition as in [2] during 10 minutes, the carboxyl group content rises to 0.15% while PD remains constant. Meanwhile according to the data in [2], during the course of cellulose oxydizing in rigid conditions (0.7% hydrogen peroxide solution, 80^0C, duration - 5 hours, pH > 11), the carboxyl group content equals only 0.06-0.08%, although PD reduces almost twice. The experimental data on the aldehyde group content in cellulose, as cited by different authors, are also contradictory.

The purpose of the present work consists of the experimental investigation and theoretical foundation of the phenomena occurring in cotton cellulose under the influence of alkali-peroxide systems.

Native cotton cellulose was used in the form of a fabric purified of admixtures in a sodium hydroxide solution. The degree of purification was varied by altering the alkali content (60-180 g/l) and by adding the reducing agent (Rongalite - 20 g/l) and the complexone (Trilon - 5 g/l). The complete removal of pectine substances and starch size was achieved by special treatments [5].

The cellulose oxidizing process was carried out at a temperature of 100^0C and at a liquor ratio of 1:1 for 1-60 minutes in the composition containing: hydrogen peroxide (100%) - 7g/l; sodium silicate (d = 1.4) - 18g/l; and sodium hydroxide (100%) - 5g/l.

Cellulose PD was calculated according to the specific viscosity of 0.1% cuprammonium cellulose solutions. Aldehyde and carboxyl group concentrations were defined using iodometric and calcium-acetate methods, respectively [5].

In our opinion, the main purpose of the discrepancy of the results on oxidized cellulose functional group content (and on the lack of correlation between the mentioned index and the decrease in PD) lies in the different degrees of the removal of aldehyde- and carboxyl-containing admixtures from cellulose during its preliminary purification and oxidizing.

There is no doubt that the high content of aldehyde (0.11%) and carboxyl (0.13%) groups in native cellulose is due to the presence of definite admixtures. Pectins and carbonic fatty acids are the main admixtures that serve as the bearers of carboxyl groups while starch size is the bearer of aldehyde groups. The content of the mentioned admixtures in native cellulose, according to different sources in the literature, is: pectins 1.0-1.2%; fatty acids 0.2-0.4%; sugars 0.25-0.30%; and starch size (for fabric) 3.0-5.0%.

It is well-known that pectins and size are removed completely with great difficulty, even in extremely rigid conditions of fiber purification and therefore special treatments are required (Table 1). But usually researchers do not control the residual content of the admixtures under discussion.

The data, presented in Fig.1 and 2, give the opportunity to estimate the dependence of the removal degree of pectin and starch size (achieved after alkaline treatment) upon the alteration of carboxyl (Fig. 1) and aldehyde (Fig. 2) group concentration in cellulose.

The difference between the degree of reduction of carboxyl group (curves 2, 4) and pectin (curves 1, 3) content during the first three minutes of treatment is apparently connected with the process of fatty acid saponification.

After five minutes of alkaline treatment, the complete coincidence of curves is observed. At the end of the process, the degree of pectin removal reaches 77%, while the carboxyl group content reduces by 73%.

The kinetic curve character (Fig.2) indicates the definite intercorrelation between the diminution of aldehyde group concentration and the degree of starch extraction. The mentioned curves are not identical because a definite part of the aldehyde groups belongs to cellulose itself. The comparison of the curves shows that such part accounts for approximately 55% of total aldehyde group content.

Table 1.

Conditions of purification		Residual content, %		
Solution composition (g/l)	τ, min	pectins	waxes	starch
Without purification	-	100	100	100
Sodium hydroxide - 60	20	29.0	60.0	8.0
Sodium hydroxide - 180, Rongalite - 20	3	47.0	50.6	10.1
Sodium hydroxide - 60, Trilon - 5	20	24.9	60.1	11.3
According to [5]	180	0	-	-
According to [5]	60	-	-	0

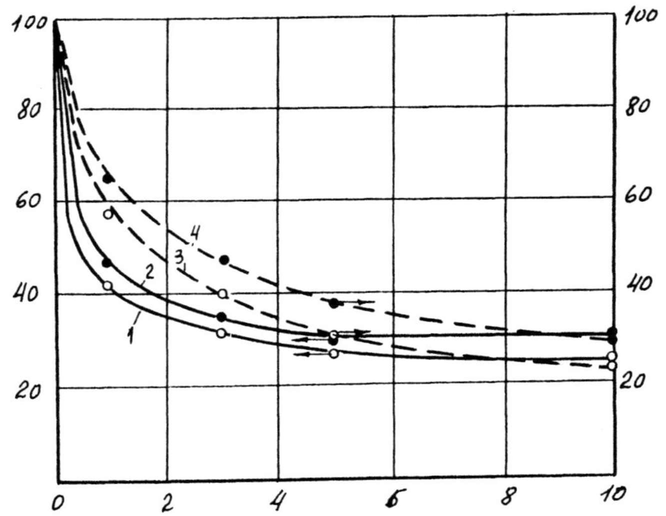

1.Kinetics of carboxyl group (1, 2) and pectin (3, 4) content alternation in cellulose during its alkaline (1, 3) and alkaline reducing (2, 4) treatments (NaOH - 180 g/l, Rongalite - 20 g/l)
Y = content of carboxyl groups, %;
Y = content of pectin, %;

X = τ, minutes.

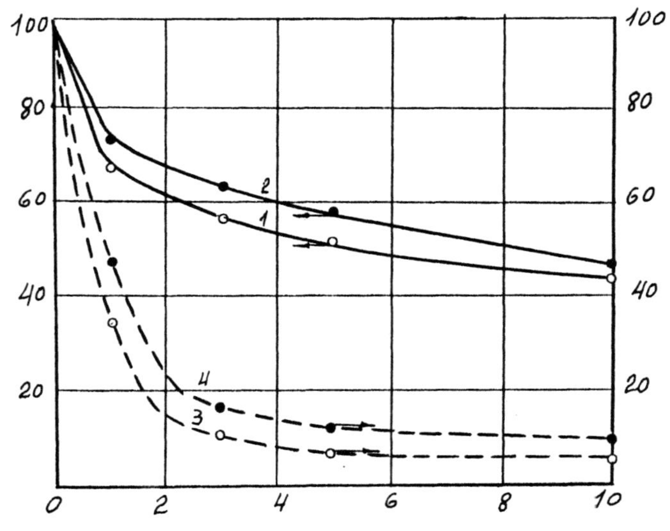

2. Kinetics of aldehyde group (1, 2) and starch size (3, 4) content alteration in cellulose during its alkaline (1, 3) and alkaline reducing (2, 4) treatments (NaOH - 180 g/l, Rongalite - 20 g/l)
Y = content of aldehyde groups, %;
Y = content of starch size, %;

X = τ, minutes.

The rise of aldehyde and carboxyl group content after the addition of the reducing agent (preventing pectin and starch oxidizing destruction) in the alkaline solution is another proof of the clear-cut correlation between the concentration of functional groups and corresponding admixtures in cellulose (Fig. 1, 2, curves 2, 4).

The results obtained demonstrate convincingly that the change in the cellulose functional group content in the course of alkaline treatments is connected with the removal of admixtures, but not with chemical transformations.

Table 2.

Conditions of purification		Duration of the oxidizing process, min	Group content, %		PD
Solution composition (g/l)	τ, min		aldehyde	carboxyl	
Without purification	-	0	0.11	0.13	2700
		10	0.078	0.098	2300
		20	0.075	0.091	2120
		30	0.069	0.087	2020
		40	0.065	0.085	2000
		60	0.064	0.085	1900
Sodium hydroxide - 60	20	0	0.073	0.030	2580
		10	0.064	-	2220
		20	0.064	0.018	-
		30	0.064	0.013	2170
		40	0.057	0.012	-
Sodium hydroxide - 180 + Rongalite - 20	3	0	0.075	0.041	2620
		10	0.065	0.027	2430
		20	0.055	0.015	2300
Sodium hydroxide -60 + Trilon -20	20	0	0.085	0.020	2500
		10	0.067	0.018	2420
		20	0.067	0.017	2360
		30	0.067	0.015	2270
		40	0.067	0.012	2200
According to [5]	-	0	-	0.005	-
		10	-	0.005	-
		30	-	0.005	-

The influence of hydrogen peroxide on the alteration of the chemical content and PD of cellulose can be illustrated by the data in Table 2. It is easily seen that in all cases, the oxidizing of cellulose causes a decrease in the amount of functional groups. The rate of this process depends upon the preliminary purification degree of native cellulose. Therefore, the aldehyde group content in native cellulose reaches a stable value of 0.064% after 60 minutes of oxidizing. For purified cellulose materials, this index is already achieved after 20 minutes. The carboxyl group concentration in native cellulose diminishes only to 0.085% after 60 minutes of oxidizing. Cellulose, having been partly purified of pectin substances, contains 0.012-0.015% of carboxyl groups after 20-30 minutes of oxidizing treatment. In completely purified cellulose, the carboxyl group concentration (equal to 0.005%) has not been changed after 30 minutes of hydrogen peroxide influence. Nevertheless, as the decrease in PD exhibits, the destruction processes in cellulose still take place.

Thus the break of β-glycoside links in cellulose is not always caused by the variation of polymer chemical composition. The phenomenon can be explained through modern theoretical concepts.

According to the new viewpoint, the primary act of cellulose chemical transformations is the alteration of the energetic state of heterocycle due to its participation in the formation of the labile intermediate complex with a nucleophilic particle [7-9]. As a result, the rupture of the β-glycoside link is observed.

In former publications [10, 11], basic demands were formulated for the nucleophilic catalysts of alkaline hydrolysis of cellulose polysugars and starch. It is probable that the perhydroxile-ion (which possesses nucleophilic properties [3]) satisfies these demands and in mild oxidizing conditions acts not as a reducing agent but as a nucleophilic catalyst of alkaline hydrolysis.

The process can be represented by the following diagram which illustrates that the break of β-glycoside links is not accompanied by the formation of new functional groups:

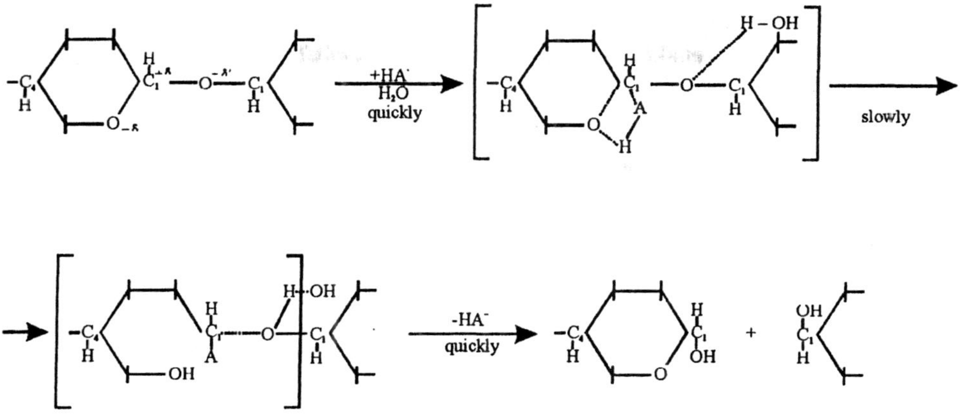

where HA = HO_2^-.

The accumulation of the discussed functional groups in cellulose is the second act of its chemical transformation. More severe time and concentration conditions of cellulose treatment with the oxidizing agent promote the interactions of the latter with the secondary reaction centers (belonging to the second, third and sixth carbon atoms) and the oxidizing of hydroxyl groups up to the aldehyde and carboxyl groups. In this case, the macromolecule rupture is accompanied by the variation in the chemical composition of cellulose.

The analysis of literary and obtained experimental data give the basis to consider that the classical theory of cellulose oxidizing destruction (the main principles of which are given in different training aids on chemical technology of fibrous materials [12-14]) does not correspond to the real conditions of cellulose textile material bleaching and therefore needs to be determined more precisely in the light of modern scientific investigations.

References

1. V.I. Ivanov, E.D. Kaverzneva and Z.I. Kuznetsova, *Rep. USSR AS,* **86** (2) (1952) 301 (in Russian).
2. V.I. Ivanov, E.D. Kaverzneva and Z.I. Kuznetsova, *Izv. USSR AS,* (2) (1953) 374 (in Russian).
3. P. Ney, *Melliand Textilberichte,* **63** (6) ( ) 443.

4. V.I. Lebedeva and T.V. Skatova, *Izv.Vuzov. Technol. Tekst. Prom.*, (4) (1978) 100 (in Russian).

5. F.I. Sadov et al., *Laboratory Practical Work in Chemical Technology of Fibrous Materials* (Moscow: Gizlegprom, 1963) (in Russian).

6. I.N. Yermolenko, *Spectroscopy in the Chemistry of Oxidized Celluloses* (Minsk, 1959) (in Russian).

7. V.I. Ivanov et al., *Rep. USSR AS,* (6) (1980) 1341 (in Russian).

8. N.Ya. Kuznetsova and V.I. Ivanov, in *Collected Articles: Chemistry and Physical Chemistry of Polyacetales,* Ed.V.A. Afanasy'ev (Frunze: Ilim,, 1974) 5 (in Russian).

9. V.E. Zhavoronkov and G.M. Korneva, in *Collected Articles: Chemistry and Physical Chemistry of Polyacetales,* Ed.V.A. Afanasy'ev (Frunze: Ilim,, 1984) 22 (in Russian).

10. E.L. Kadykova, S.M. Gubina and B.N. Mel'nikov, *Izv. Vuzov. Khimiya i Khim. Tekhnol.,* (11) (1991) 88 (in Russian).

11. V.G. Stokozenko, E.L. Kadykova and S.M. Gubina, *Khimiya Drevesiny* (3) (1992) 33 (in Russian).

12. F.I. Sadov, M.V. Korchagin and A.I. Matetskiy, *Chemical Technology of Fibrous Materials:Training Aids for the Institutes* (Moscow: Lyogkaya Promyshlennost, 1968) (in Russian).

13. B.N. Mel'nikov, T.D. Zakharova and M.N. Kirillova, *PhysicalChemistry Bases of the Finishing Industry Processes* (Moscow: Lyogkaya i Pishchevaya Promyshlennost, 1982) (in Russian).

14. G.E. Krichevskiy, M.V. Korchagin and A.V. Senakhov, *Chemical Technology of Textile Materials* (Moscow: Legprombytizdat, 1985) (in Russian).

The Institute of Chemistry of Non-aqueous Solutions of the Russian Academy of Sciences, Ivanovo

A NEW APPROACH TO THE INTENSIFICATION OF CONTINUOUS PROCESSES FOR COTTON FABRIC DYEING USING WATER-SOLUBLE DYES

Ju.A.Kalinnikov

The relationship between the state of water-soluble dyes in solution and their adsorption by cellulose is discussed. A new approach to the intensification of cotton fabric dyeing with water-soluble dyes is formulated and scientifically grounded. The approach consists in stimulating the association of dyes in the solutions formed in the pores of the fiber. New dyeing methods created on the basis of the proposed approach are considered.

The less developed part in the modern theory of dyeing concerns the mass-transfer of dyes from the solution into the fibrous polymer. Specifically, the relationship between the state of dyes in solution and the processes of their adsorption and diffusion in the fiber has been unsatisfactorily studied. However, the opportunity exists to enhance the effectiveness of dye adsorption by altering dye solutions.

In publications devoted to this problem, the role of the dissolved dye fraction in the process of dye transfer from the solution to the fiber in some classic technologies has been only partly determined. Large scale generalizations are absent in the literature - such ones that can serve as a theoretical basis for the creation of a series of radically new dyeing technologies. At present, the opinion has taken root that the alteration of the state of dyes in solution as a device of dyeing intensification is almost exhausted and can neither produce new technical effects nor raise the qualitative level of dyeing processes. This situation is an obstacle to the development of theoretical and practical textile material dyeing.

One possible way out of this dilemma lies in looking again at the essence of those processes, taking place in dye solutions, from the position of a general fundamental theory of solutions. The fruitfulness of such an approach is illustrated by the research results of cotton fabric dyeing with active and vat dyes.

To realize this intention, four liquid media were chosen: liquid ammonia, its aqueous solutions, aqueous-electrolyte systems, and aqueous-ammonia-electrolyte systems. Liquid ammonia has been studied thoroughly as an individual solvent and as an addition to water [1, 2]. Furthermore, this solvent is known as an extremely strong means for cellulose modification in different methods of dyeing and of finishing textiles [3, 4]. Under the influence of liquid ammonia, cellulose becomes mercerized and swells several times quicker than in the solutions of sodium and potassium hydroxides [3]. Hence, the correlation between the state of dyes in solution and their adsorption in the course of dyeing becomes redoubled, and this fact strengthens the conclusiveness of the ideas that will be developed in this article.

Ionic processes in aqueous and mixed aqueous-nonaqueous solutions of inorganic salts have been also studied in detail [5].

Thus, the theoretical background for the development of this scientific and practical trend of investigation is available.

Small additions of ammonia to the aqueous solutions of dyes change such parameters as the pH-values and the structure of the solution which (according to the theory of solutions) determine interparticle interactions in liquid media [6]. The interparticle interactions, in their turn, move the interphase transfer of dye from the solution into the fiber in one direction or the other. It was planned to influence subtly these parameters, using not only ammonia but also inorganic salts. The latter are considered to be the recognized regulators of ionic dye association in water [7, 8]. The aim was to change the state of dyes in solution in such a way that the dye would be absorbed by the polymer more intensely.

It is accepted that inorganic salts reduce electrostatic repulsion between the dye and the cellulose and promote the association of dyes in solution. While the reduction of electrostatic repulsion is unreservedly positive from the standpoint of the intensification of dyeing, the dye stimulation promotion is taken as an unavoidable consequence of salt application. That is why attempts are usually undertaken to restrict the enhancement of dye association. The harmfulness of dye association in the dyeing bath consists of an extremely active adsorption of the associates on the fiber surface, resulting in weak dyeing fastness. Hence, dye association in solution is considered to be a negative phenomenon.

Comprehensive investigations into the properties of aqueous and ammonia dye solutions (as well as studies into the correlation between the state of dye in solution and dye adsorption and diffusion in cellulose) give the opportunity to convert the process of dye association into the key factor of the intensification of cotton fabric dyeing. The first stage of our research has been devoted to the study of continuous dyeing technologies in which the operation of fabric impregnation in the dyeing solution occurs just after soaking the fabric in liquid ammonia (without removing the ammonia from the fabric) and in which the composition of the aqueous dyeing bath stays traditional. Such organization of the technological process is deemed to be optimal [3].

At first by mathematical modeling, the consecutive treatment of a capillary-porous body in liquid ammonia and water [9], and secondly, experimentally by using fine-porous glass bodies [10], it was proved that the just mentioned sequence of treatments leads to the formation of highly concentrated ammonia solutions in the internal volume of solid porous substrates. These solutions preserve the media for dye adsorption by the fiber.

Then by means of conductometric and spectrophotometric methods, it was shown (both qualitatively and quantitatively) that the tendency of active dyes to associate in liquid ammonia and aqueous ammonia is much greater than in water. It was also established that the well-known stimulating action of inorganic salts on dye association in water is displayed more strongly while applying liquid and aqueous ammonia. From this standpoint, sodium and potassium nitrates are the most effective [4].

These ideas are illustrated convincingly through the electronic spectra of the reactive dye C.I. Reactive Red 45 in the solutions under discussion. This dye was used as an instrument of investigation of dye state in the solutions formed during technological operations because this dye association picture has been preliminarily identified [11].

The spectra of C.I. Reactive Red 45 in pure water (curve 1) and in water with the additions of sodium nitrate (curves 2-8) are presented in Fig.1. Some lowering of light adsorption of the dye, as the salt concentration in aqueous dye solution is increased (it is

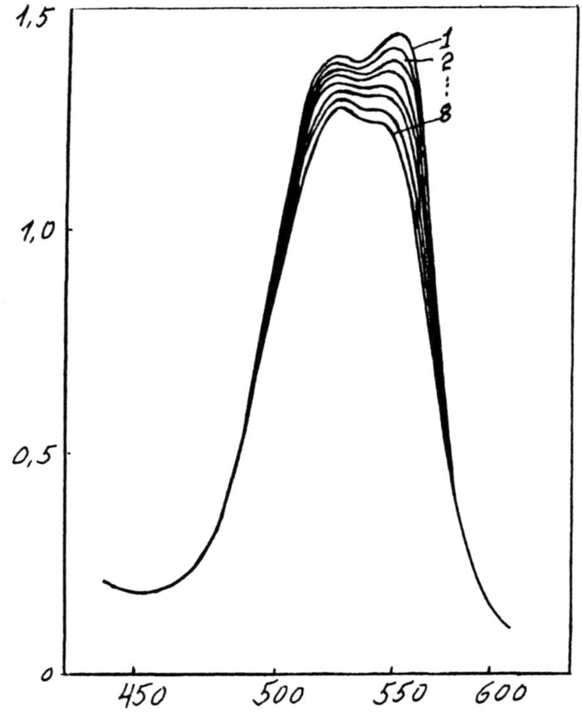

Fig.1. Electronic light-absorption spectra of the dye C.I. Reactive Red 45 in pure water (1) and in aqueous solutions of $NaNO_3$ (2-8) with a salt concentration (g/l) of 2.0; 5.0; 10.0; 20.0; 30.0; 40.0; 50.0, respectively. Dye concentration = 0.5 g/l.

X = λ, nm;

Y = A (light absorption = optical density of the solution).

expressed more strongly on the long-wave peak), indicates the appearance of small dye associates in aqueous salt solutions. On the contrary, in the aqueous ammonia - sodium nitrate system, there is a spectral evidence of violent reactive dye association (Fig.2). The proves are: a sharp decrease in light adsorption on the long-wave band until almost complete disappearance and the formation of a new short-wave band, predominating in solutions with a high salt concentration and signaling the presence of large dye associates [11].

These facts testify to the important role of dye association in solution during fabric dyeing according to the cited technological pattern. Analyzing the totality of the results

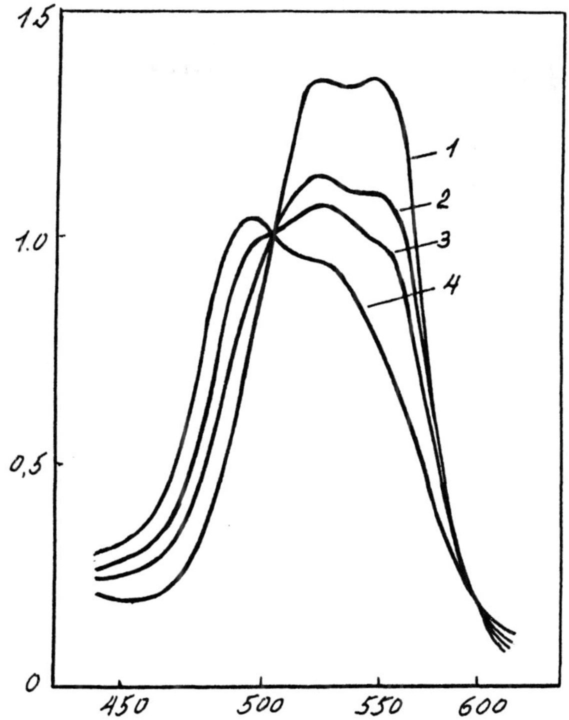

Fig.2. Electronic light-absorption spectra of the dye C.I. Reactive Red 45 in aqueous ammonia (1) and in aqueous-ammonia solutions of $NaNO_3$ (2-4) with a salt concentration (g/l) of 2.0; 5.0; 10.0, respectively. Dye concentration = 0.5 g/l..

X = λ, nm;

Y = A (light absorption = optical density of the solution).

of our investigations into reactive dye solutions and applying our results to continuous ammonia-using processes, one can see that the most potent dye association takes place in liquid media formed after two consecutive soaking operations: first in liquid ammonia and then in an aqueous dyeing bath.

Taking into account the examined phenomenon of strong ionic dye association in liquid ammonia, the well-known fact of the intense stimulating effect of ammonia on dye sorption and diffusion in cellulosic textiles [3] may be interpreted as follows: Dye associates, formed rapidly in the fiber pores when the fabric is impregnated, are characterized by a reduced dye affinity to the solvent and a heightened dye affinity to cellulose in comparison with monomers. Dye transfer from the liquid phase into the fiber is accompanied by the local decrease of dye concentration in the internal solution in pores and, consequently, by the appearance of a concentration gradient between the solutions inside and outside the polymeric substrate. The concentration gradient is a motive force of dye diffusion into the textile material. In the liquid medium inside the fiber, the dye particles enlarge to such extent that they become unable to stay in solution. They are adsorbed on the inner surface of the polymer and, as a result, provide an inflow of new portions of dye to the fixation zone. Thus, the tendency of the whole system to reach thermodynamic equilibrium conditions the chain character of the processes under discussion and therefore the effective dye adsorption in the fiber. The dimensions of the pores define the associate maximal dimensions and finally the dimensions of dye particles that form the fabric coloring.

The discussed mechanism of intensification deals with such free-space elements of fiber and fabric in which the steric factor does not limit dye molecule association. This is a general case, but this does not mean that the described approach can promote dye adsorption only inside the mentioned elements of textile material. It is evident that in similar conditions, dye adsorption will be more intense in the pores of every size - even those that can accommodate only one dye molecule, because the interparticle interactions causing association and adsorption are exactly the same [8, 9]. That is why dye adsorption by the fiber will be more effective if a liquid medium favoring association is used. The enlargement of dye particles in the solution is not a necessary condition for stimulation of the adsorption.

All the above makes it obvious that in order to work out intensified continuous methods of cotton fabric dyeing with ionic dyes, it is indispensable to organize the whole process in such a way that dye association will be an incentive not a hindrance to adsorption. The principle of building such processes is also clear: it is advisable to stimulate dye association in the liquid medium formed in the internal fiber and fabric volume during soaking operations. The larger the difference in dye tendency associating in the internal solution in comparison with the external one, the more convincing is the result.

These conditions are fulfilled when sodium or potassium nitrate is added in the dyeing bath because these salts intensify ionic dye association in aqueous ammonia more strongly than others [4, 12, 13].

The approach that has been formulated is realized in several radically new, intensive cotton fabric dyeing processes based on the application of liquid ammonia and the use of the thermal method of reactive dye fixation (as the most progressive at present) [14, 15]. The practical evidence of the adduced theoretical conception is illustrated by the data presented in Table 1 where the indices of calico dyeing with the reactive dye C.I. Reactive Red according to our worked out "ammonia-electrolyte" method (and the most close to it "ammonia" process [3], created without taking dye association into account) are given.

The results show that the embodiment of the proposed approach considerably intensifies dye adsorption and fixation. The increase in dye fixation is caused not only by the activation of dye mass-transfer into the polymer but also by the change in chemical reactions of the formation of colorings. This change is discussed in detail in [12].

Table 1

Dyeing process	Dye content in the fiber, g/kg		Degree of dye fixation, %
	adsorbed	fixed	
"Ammonia"	14.3	12.1	82.7
"Ammonia-electrolyte"	18.4	17.7	95.6

The possibility to apply a specific theory to other objects and systems is the convincing proof of the validity of each theory. A positive result will outline the borders of the theory's applicability and reveal its capacity to predict. Considering these reasons, we have made an attempt to use our newly developed approach in order to create the combined technology of cotton fabric mercerizing and dyeing with vat dyes in an aqueous medium. For our purposes, the most suitable framework is that of the traditional suspension dyeing method. It is advisable to stimulate dye adsorption by means of modifying the stages of fabric soaking in the dye suspension and the subsequent treatment of fabric in the alkaline reducing solution. These are the main operations determining the quantitative aspect of dye mass-transfer into the solid textile polymer phase. The remaining stages of this technology are responsible mostly for the quality of colorings.

The purpose of soaking is to bring the solid vat dye particles onto the surface of fibrous material effectively and evenly. The content of the adherenced dye determines the content of fixed dye to a considerable extent.

During the alkaline reducing treatment of the fabric, the vat dye is converted into a water-soluble form, and as a result, it diffuses inside the substrate together with the solvent and is adsorbed on the inner surface of the cellulosic fiber in conformity with the mechanism peculiar to ionic dye adsorption. Up to now, much attention has been paid to the exploration and search for means to intensify the chemical reaction of converting vat dyes from an oxidized form into a reduced one, and no attempt has been undertaken to diminish the losses of dissolved dyes due to their diffusion from the material back into the alkaline reducing bath.

The analysis of the physico-chemical essence of the processes taking place during fabric soaking and treating in the alkaline reducing solution has revealed the suitability of the principle of the construction of continuous technologies (the processes of fabric dyeing with reactive dyes served as the object of our investigation) to dyeing processes in which vat dyes are used. These dyes are chemically unchanged under the influence of sodium hydroxide, so the latter may be added to the suspension in a concentration sufficient enough to make cellulosic fiber mercerized. Moreover, a mercerizing treatment combined with a dyeing operation will raise the quantity of dye brought into the substrate. Starting from these conclusions (concerning the intensification of dye transfer from the solution into the fiber by means of purposeful stimulation of dye association), it is necessary to bring to the cellulose surface such a substance which can be easily dissolved in the alkaline reducing solution and therefore provide a considerable difference in liquid medium compositions inside and outside the porous fabric. It is essential that the liquid in the inner volume must stimulate the association of dye to a much higher degree than an alkaline reducing solution located in the bath. In this case, the adsorption of dye will be more effective in accordance with the mechanism discussed earlier for reactive dyes. The distinction consists in the fact that, in the case of reactive dyes, the association stimulates dye penetration into the fiber while in the case of vat dyes the deliberately caused association prevents the reverse diffusion of dye from the internal liquid phase into the external solution.

On the basis of the results from our research into reactive dye solutions, three salts, $NaNO_3$, KNO_3, and NaCl, were chosen to cause the association of vat dyes. These salts satisfy all the aforementioned demands for substances that can be used to intensify cotton fabric dyeing with suspensions of vat dyes and to create a new combined process of cotton fabric mercerizing and dyeing.

Experimental evidence of the association of vat dyes in the alkaline reducing solution under the influence of inorganic salts is needed to confirm the correctness of the proposed precondition. A spectrophotometric method was used to study the solutions of vat dyes. The most complicated problem in this study was the achievement of stability of the initial spectrum of the reduced form of vat dye. This was solved by means of a thorough selection of concentrations for all components in the mixture.

Fig.3 reflects the influence of NaCl additions on the state of vat dye C.I. Vat Orange 3 (59300) in the alkaline reducing solution. The alterations are limited to the consistent lowering of the intensity of light-absorbing bands as the salt content in the solution is increased and displays the typical spectral picture of the association. The identical tendency manifests itself when one of the other cited salts, or sodium hydroxide, or simultaneously both of them are introduced into the solution of a vat dye. The difference in the effect produced by different substances lies in the degree of dye association. This degree can be evaluated provided that the data are represented as curves in the coordinates where Y = intensity of light absorption and where X = the concentration of inorganic electrolyte (Fig.4).

As is seen in Fig.4, the association of the vat dye in a solution containing sodium hydroxide is minimal among the investigated solutions (curve 1). The associative interactions between the dye molecules are expressed more violently in the presence of the inorganic salt (curve 2). Such mutual arrangement of the curves may be explained by the less salting-out effect of NaOH in comparison with NaCl and by the action of sodium hydroxide as a chemical reagent, which converts a vat dye from the form of a vat-acid, hardly dissolving in water, into the well-dissolving form of its sodium salt. The abrupt reduction of the intensity of light-absorption takes place when sodium chloride is added to the dye solutions containing sodium hydroxide in concentrations of 150 g/l and 180 g/l (curves 3 and 4, correspondingly). There is no doubt that the process of the association of vat dye in systems where the salt and the hydroxide are present simultaneously is much more pronounced than in those systems with equal concentrations of NaCl or NaOH taken separately. This leads to the conclusion that the creation of a hydroxide and salt-containing solution (stimulating the association of vat dyes greatly) in the internal volume of textile material will provide enhanced dye adsorption on the internal surface of a cellulosic fibrous substrate.

The application of solutions, containing both hydroxide and salt, for the preparation of the suspensions of vat dyes and the usage of definite parameters for fabric drying (following the soaking stage) are the principal peculiarities of the newly developed process. These operations ensure not only cotton fabric mercerization and attracting the dye to the fabric surface, but also the effective resultant adsorption of the reduced form of vat dye by the fiber. The process conditions of this dyeing method are:

1. Impregnation of fabric with the suspension, g/l:

vat dye	20 - 30
sodium hydroxide	140 - 170
sodium chloride	20 - 60.

2. Drying of the material at a temperature of 100-160^{0}C for 30-55 sec.
3. Soaking of fabric in the aqueous solution of sodium hydrosulphite (40 - 60 g/l) at a temperature of 20 - 25^{0}C for 2 - 5 sec.
4. Steaming of fabric by saturated water steam at 100^{0}C for 30-60 sec.
5. Oxidizing, washing, and drying the fabric according to the procedure adopted or methods of dyeing from suspensions.

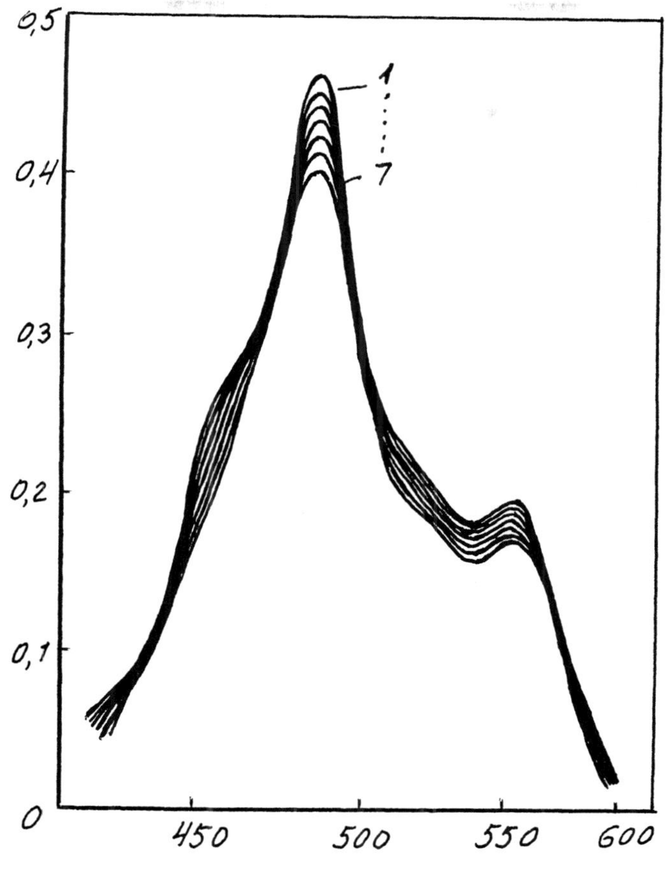

Fig.3. Influence of NaCl concentration on the electronic light-absorption spectra of the dye C.I. Vat Orange 3 in the alkaline -reducing solution (g/l): 1 - 0; 2 - 10; 3 - 30; 4 - 50; 5 - 70; 6 - 80; 7 - 110.

X = λ, nm;

Y = A (light absorption = optical density of the solution).

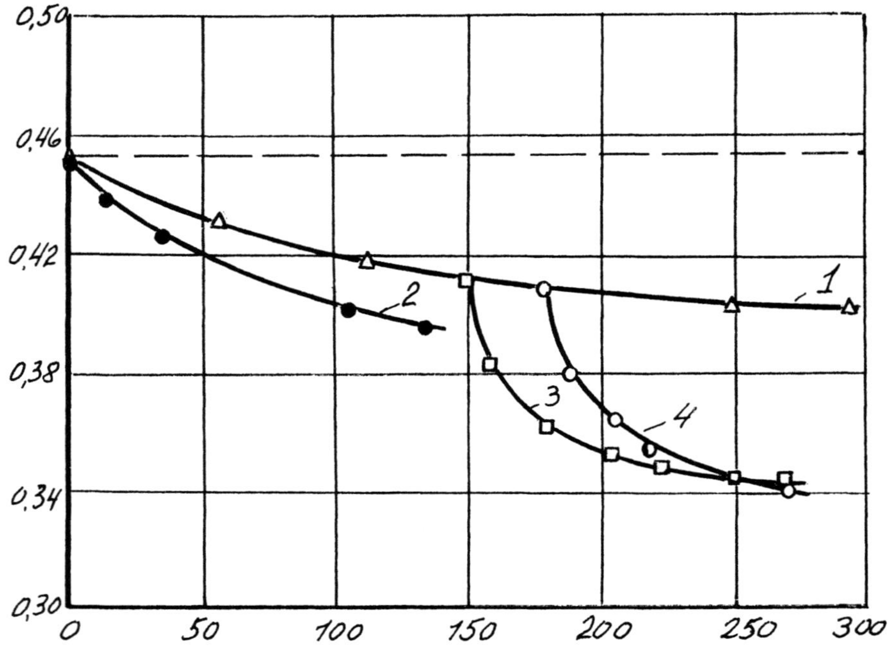

Fig.4. Influence of the concentration of inorganic electrolytes on the electronic light-absorption spectra of the dye C.I. Vat Orange 3: 1 - NaCl; 2 - NaOH;
3 - 150 g/l NaOH + NaCl; 4 - 180 g/l NaOH + NaCl.
X = *c*, g/l
Y = A (light absorption = optical density of the solution).

The results show that the embodiment of the proposed approach considerably intensifies dye adsorption and fixation. The increase in dye fixation is caused not only by the activation of dye mass-transfer into the polymer but also by the change in chemical reactions of the formation of colorings. This change is discussed in detail in [12].

Table 2.

Dye (C.I.)	Dye content in the fiber, %	Dye fixation degree, %	Ratio of the light reflection coefficients of right and left sides of the fabric	Steadiness of colorings, number	
				to dry rubbing	to wet rubbing
Vat Green 1 (59825)	19.0 / 19.4	82 / 91	0.84 / 0.97	4 / 4	3 / 4
Vat Orange 3 (59300)	21.4 / 22.4	89 / 95	0.81 / 1.00	4 / 4	4 / 4
Vat Violet 1 (60010)	30.5 / 39.00	95 / 97	0.97 / 1.00	4 / 4	3 / 4

This new combined process is characterized by the data in Table 2, where the technical indices of the colorings of serge (article 3217) by vat dyes according to traditional (numerator) and proposed (denominator) methods are listed as conventional ratios.

It can be easily seen that the developed technology in comparison with the traditional one permits (together with an increase in dye content in the fiber):

- a 2 - 9 % heightening of the dye fixation degree;
- the achievement of more even dye distribution in the fibrous polymer;
- the improvement of fastness of colorings to wet rubbing.

To sum up the results of our research, it may be concluded that our developed ideas were both theoretically and practically confirmed. This means that a new approach to the intensification of continuous processes for cotton fabric dyeing has been created. Such an approach presupposes a deliberate change in the state of a dye in solution.

References

1. G. Tompson, *Electrons in Liquid Ammonia* (Moscow: Mir, 1979) (Russian Edition).
2. A.P. Kreshkov, *The Analytical Chemistry of Non-aqueous Solutions* (Moscow: Khimiya, 1982) (in Russian).
3. B.N. Mel'nikov , M.N.Kirillova and A.P. Moryganov, *The Present Situation and the Possibilities for the Development of Textile Material Dyeing Processes* (Moscow: Lyogkaya i pishchevaya promyshlennost, 1983) (in Russian).
4. B.N. Mel'nikov, A.P. Moryganov and Yu.A. Kalinnikov, *The Theory and Practice of Intensive Fixation of Dyes on Textile Materials* (Moscow: Legprombytizdat, 1984) (in Russian)
5. G.A. Krestov, The Thermodynamics of Ionic Processes (Leningrad: Khimiya, 1984) (in Russian).
6. G. Gordon, The Organic Chemistry of Electrolyte Solutions (Moscow: Mir, 1979) (Russian Edition).
7. T. Vickerstaff, The Physical Chemistry of Dyeing (Moscow: Gizlegprom, 1956) (Russian Edition).
8. G.E Krichevskiy et al., *The Chemical Technology of Textile Materials* (Moscow: Legprombytizdat, 1985) (in Russian).
9. F.Yu. Telegin et al., *Izv.Vuzov. Tekhnol. Tekstil. Prom.*, (5) (1989) 69 (in Russian).
10. I.Yu. Vashurina et al., *Izv.Vuzov. Tekhnol. Tekstil. Prom.*, (2) (1991) 57 (in Russian).
11. A.N. Terenin, *The Photonics of Dye Molecules* (Leningrad: Nauka, 1967) (in Russian).
12. G.A. Krestov et al., *Non-aqueous Solutions in Engineering and Technology* (Moscow: Nauka, 1991) (in Russian).
13. I.Yu. Vashurina, *Reactive Dye Association and the Creation of Effective Cotton Fabric Dyeing and Finishing Technology*, (Dissertation for the degree of Candidate of technical science, Ivanovo, 1988) (in Russian).
14. Patent Certificate 1168639, USSR. Registry of Inventions 1985, Number 27.
15. Patent Certificate 1359380, USSR. Registry of Inventions 1987, Number 46.
16. Patent Certificate 1523603, USSR. Registry of Inventions 1989, Number 43.

The Institute of Chemistry of Non-Aqueous Solutions
of the Russian Academy of Sciences, Ivanovo

Part IV

New Processes and New Equipment for Textile Material Finishing

THEORY AND PRACTICE OF COMPOUND CHEMICALS DEVELOPMENT IN TEXTILE INDUSTRY

A. P. Moryganov

The review reports on general principles of development of compound chemicals with desired functional properties intended for intensification of textile finishing. The emphasis is put on the chemicals which affect the state of liquid-phase systems. Examples of both the chemicals developed and their industrial application in textile industry are given.

Any progress in equipment and textile finishing technology is impossible without highly efficient compound chemicals which make it possible to impart desired properties to fibers provided both full utilization of the chemicals and lower levels of environment contamination are realized. A number of great European companies (Hoechst, BASF, Boeme AG, etc.) are producing a wide set of chemicals for intensification of textile treatment. However, in Russia only simple chemicals are broadly used, such as rongalite, urea, wetting agents, but compound ones are rare nowadays.

During the 1980s complex investigations have been carried out in the Institute of Non-Aqueous Solutions Chemistry of the Russ. Academy of Sciences (INASC). They have been dedicated to fiber-forming polymers plasticization, state of dyes in solutions and melts, mechanisms of interaction between compound chemicals and fibers both under normal conditions and due to activation by non-conventional, i. e. non-physicochemical, methods. These investigations have made it possible to work out new approaches to development of liquid-phase chemicals with desired properties aimed at intensification of preparation, dying, printing, and finishing of fibrous materials [1]. Routs to the new compound chemicals proposed by the INASC are as follows:

- to raise efficiency of individual plasticizers and to search for co-reagents which affect different structural elements of fiber-forming polymers, the co-reagents being intended for fibers activation in dyeing, printing, linear dimensions stabilization;

- to search for intensifiers, catalyzers, and compositions of ingredients which supplement one another, the reagents being intended for catalyzing of chemical reactions in textile preparation and coloration;

- to activate the liquid-phase systems by many-sided action (physicochemically, mechanochemically, magnetically).

One of the examples illustrating realization of the first direction is an intensifier Lakran based on potassium thiocyanate and urea. Lakran is intended for fixation of reactive dyes and mixtures of reactive and disperse dyes on cellulose materials. The main points to develop it on the base of this binary mixture have been considered as follows [2, 3]. On the one hand, plasticizing synergistic effect of these compounds on cellulose makes it possible to reduce the polymer glass transition by 25-30°C in the process of heat treatment. The synergism being due to combination of the molecular and structural mechanisms raises sharply the fiber free volume and hence dyes fixation rate. On the other hand, urea stability to thermal destruction increases significantly because of a change in intermolecular interactions in the mixed melt.

Temperature range of the plasticizer stable melt existence increases from 30-40°C to 138-140°C in comparison with pure urea (Fig. 1). It is noteworthy that isocyanic acid generated by decomposition of urea neutralizes the basic component. Thus, the dyeing bath pH value decreases (Fig. 2, graphs 3 and 4) and hampers fixation of reactive dyes by a fiber. The mixed intensifier consisting of potassium thiocyanate and urea (Fig. 2, graph 2) makes it possible to maintain the dyeing bath pH value at the level desired (9.8-10.3) within broad temperature range (140-200°C). That makes reactive dyes fixation more complete and therefore improves dyeing quality.

Industrial testing of Lakran has confirmed (Table 1) that high coloring intensity can be achieved by applying less quantities of printing dyes than earlier because dyes fixation increases by 10-26%. The intensifier developed is universal considering both its fixation method (hot air, saturated steam) and dyes classes (reactive, vat, disperse ones and their mixtures). It is noteworthy that application of Lakran makes it possible to give up costly ingredients of printing compositions, e. g. xylitane and glycerin, in printing cotton fabrics by the rongalite-potash method. Successful dyes fixation can be achieved not only by saturated, but overheated steam as well.

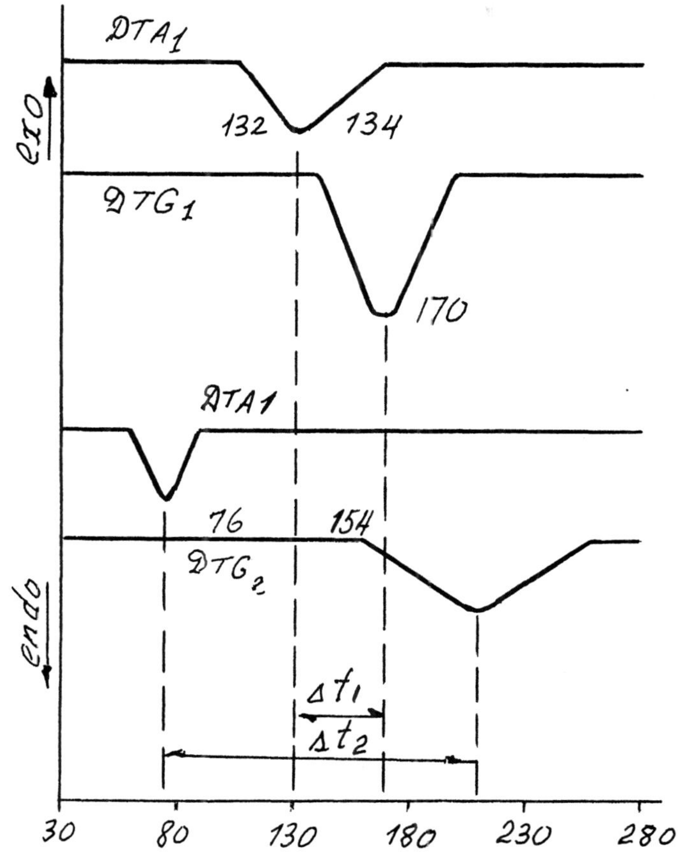

x = temperature, °C

Fig 1. Thermal analysis of urea (1) and the mixture of urea with potassium thiocyanate(2). Differences ..t_1 and ..t_2 are the temperature ranges of stable melt existence for urea and the binary mixture, correspondingly

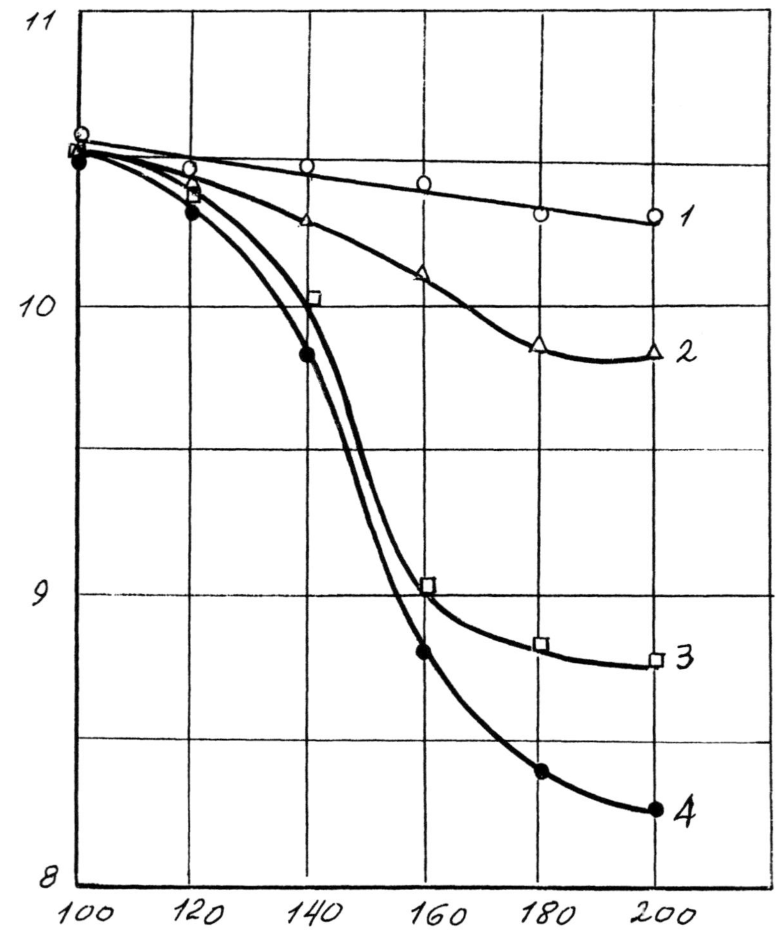

y = pH of the dyeing bath
x = temperature, °C

Fig 2. Influence of temperature and plasticizers on the dyeing bath pH value: 1. sodium carbonate (1.5%); 2. sodium carbonate (1.5%), urea (5%) and potassium thiocyanate (5%); 3. sodium carbonate (1.5%) and urea (5%); 4. sodium carbonate (1.5%) and urea (10%)

Table 1.
Results of the industrial testing of Lakran

Dye* (fabric)	Intensifires**	Quantity applied, g/kg	Fixation conditions			
			Saturated steam		Overheated steam	
			Dye quantity, g/kg	Fixation extent, %	Dye quantity, g/kg	Fixation extent, %
Reactive Green 4	1	56.6	40.5	72	39.6	70
Reactive Blue 23T						
(Satin)	2	48.8	43.1	88	42.3	86
Reactive Red 6	1	61.4	31.2	51	35.1	57
(Satin)	2	46.5	27.2	59	40.4	87
Reactive Red 6	1	34.0	24.8	73	23.8	70
Ostazine Orange						
H2R	2	32.4	26.7	82	27.4	85
(Calico)	1	36.0	-	-	23.5	65
Reactive Green 4	2	31.0	-	-	26.8	86
(Satin)	1	40.1	30.5	76	28.2	70
Reactive Green 4	2	36.4	31.1	85	29.7	82
(Satin)	1	38.8	29.7	77	27.1	70
Reactive Green 4	2	33.8	28.1	83	26.5	78
(Satin)						

* - here and afterwards the titles of the dyes are given in Russian version;
** : 1 - urea, 100-150 g/kg; 2 - Lakran, 80-100 g/kg.

Lakran being an powerful tool to improve textile coloration economically and ecologically is being manufactured by Iv'khimprom Co. (Ivanovo).

Another example of the compound chemical development in accordance with the first direction is illustrated by Table 2 [4]. Water-solvent mixing heats being characteristics of a weakening of intermolecular interactions in solutions correlate well with deviation from additivity of the fibrous material glass transition point. Application of the thermodynamic criterion, when aqueous-organic mixtures with high non-additive effect are being searched for, has made it possible to forecast high efficiency of water - triacetin mixture as plasticizer of cellulose triacetate fibers. Considering ecological harmlessness of triacetin in contrast with other organic solvents, new recommendations have been worked out to apply the substance in a) periodic aqueous dyeing of di- and triacetate fabrics, b) fixation of disperse dyes on printed linen by steam, c) thermofixation dyeing of mixed cellulose acetate - polyester fabrics. In the last case, controlled thermoplasticizing activation of fibers with considerably different sorption activities (cellulose triacetate and polyester) makes it possible - due to supplementary plasticization of cellulose triacetate - to impart the fibers nearly equal affinity to disperse dyes. The distribution coefficient values for dyes distribution between the fibrous components tend to one (Table 3). This phenomenon prevents different tints of the fibrous components which were observed earlier in coloration by conventional methods.

Table 2.
Relationship between components mixing heat and plasticizing activity of the aqueous-organic mixtures for cellulose triacetate fiber

Solvent (mole fraction)	Enthalpy, kJ/mole	t_g, ^0C
Propanol	0.42	10
Butanol	1.74	24
Benzyl alcohol	2.92	73
Triacetin	2.75	64

Table 3.
Fixation of Disperse Ruby on cellulose triacetate-polyester fabrics

Intensifier (concentration, g/l)	Hot air, 200^{0}C, 1 min				Overheated steam, 170^{0}C 4 min			
	«Melody»		«Gull»		«Melody»		«Gull»	
	K_R^b	FE^c	K_R^b	FE^c	K_R^b	FE^c	K_R^b	FE^c
Urea (100)	0.72	80	0.79	83	0.76	82	0.71	84
Urea (50) and triacetin (20)	0.93	92	0.89	90	0.96	91	0.88	90

Note: The fabric «Melody» consists of 69% cellulose diacetate and 31% polyester; the fabric «Gall» consists of 65% cellulose triacetate and 35% polyester; K_R^b - coefficient of dye distribution between cellulose and synthetic components; FE^c -fixation extent, %.

Thus, application of the compound chemicals based on triacetin makes in possible a) to shorten dyeing time, b) to reduce fixation temperature and expenditure of disperse dyes, c) to improve uniformity of mixed fabrics dyeing.

Another polyfunctional reagent being developed nowadays in the INASC is intended for periodic dyeing of polyester fibers. Table 4 compares it with a known dyeing composition for the high-temperature Rapid-Color method. The latter consists of six monofunctional auxiliaries which are costly chemicals imported. The new composition consists of a single component working as a complex one being intensifier and leveler simultaneously. It raises the dyes yield into a fiber 1.5 times and reduces oligomers concentration on a surface of yarns 7 times. Hence the yarns are not necessary to be treated with alkaline reducing agents. The new reagent provides very high levelling of color and its high fastness to dry and wet rubbing as well as raises fabrics hygroscopicity without deterioration of strength and frictional parameters [6].

The second direction of new chemicals development is based on investigation of intimate mechanisms of physicochemical processes taking place in a heterogeneous system "fiber - admixture - auxiliary agent - solvent" when a textile material is being prepared and colorated. It is necessary to understand theoretical foundations for development of intensifiers and catalyzers which supplement one another.

In the frameworks of this direction, investigations on kinetics and thermodynamics of starch hydrolysis are being based on both new experiments and literature data analysis. They have made it possible to propose a scheme of alkaline hydrolysis of starch [7]. It has been proved that alkaline hydrolysis is a catalytic process promoted by nucleophylic particles (anions), the size and dipole moment value determining catalytic activity of the particles. The hydrolysis mechanism is not only of theoretical but practical importance as well. Considering the data obtained, a new reagent consisting of sodium hydroxide and sulfide has been suggested as a desizing agent. A cold desizing of cotton fabrics has been developed [8]. A total removing of starch makes it possible to replace an inefficient treatment of technical fabrics in boilers with permanent treatment. It has raised labor productivity 11.3 times and reduced expenses 1.9 times [9].

Table 4.
Comparison of Textured Polyester Fibers Dyeing with a Conventional Composition and a New one by "Rapid-Color" Method

Dyeing composition (percentage of fiber)	Dye quantity applied g/kg	Oligomer content on a fiber surface, %	Color evenness[a], grey scale number[b]	Color fastness to rubbing	
				Dry	Wet
Conventional					
Disperse Brilliant Pink (3) Dispersogene R (1) Polyron 1.004, buffer (1) Acetic acid, pH 4.5 (1) Remole NT, carrier (1) Egonal SME, leveler (1) Leomin UR, olygomers disperser (1)	19.9	0.14	0.5	5	5
New					
DispersBrilliant Pink (3) Intensifier-leveler (7.5 g/l)	29.2	0.02	0.5	5	5

[a] Color evenness is evaluated as a tint differences after unrolling of three-layered packing; [b] Fibers with color non-evenness of not more than 2.5 % are supposed to be of the first grade

Note: the titles of the dyes are given in Russian variant

Investigations of various factors influence on thermostability and reducing activity of rongalite have been carried out in the INASC . They have shown how to adjust these properties by means of stabilizers and promoters. Application of these ingredients has made it possible to create a number of highly efficient chemicals for preparation and coloration of fabrics.

Technologies of the chemicals synthesis developed by Iv'khimprom Co. afford sharp reduction of rongalite waste as well as high stability and convenient ready forms of the compound chemicals.

One of the new rongalite-based chemicals is Rontex which, for the one hand, affords desired completeness and high rate of admixtures extraction from cellulose and, for the other hand, inhibits destruction of the polymer [10]. Rontex is being produced by Iv'khimprom Co. from 1990. It was being tested on textile factories of Iv'tex Co.

(Ivanovo). The results of testing show that Rontex is versatile enough a) to control preparation of fabrics with various functions, b) to shorten a boiling stage 1.5-2.0 times, c) to give up a stage of acid treatment. Application of Rontex provides high quality preparation, high extent of cellulose preservation, and high dyeability of fabrics with dyes of different classes. The dyeability level increases by 10-40% accompanied by uniformity and fastness of color by 0.5-1.0 units. New boiling compositions consist only of alkali, Rontex, and a wetting agent. They are easy to prepare on chemical stations of textile shops. When Rontex is being used, expenditure of water, steam, and electricity drops by 30%.

Table 5.
Rontex Efficiency for Cotton Fabrics Boiling

Boiling composition	Boiling time, min	Admixtures removed	Capillarity, mm/h	Whiteness %	Specific viscosity	Specific Dyeability %
Conventional	120	62.0	120-130	82-83	1.5	100
With rontex	60	72.0	150-180	82-87	1.68-1.87	Vat: 117-131 Reactive: 110-115 Cold: 123-142 Sulfur: 120-122

Another rongalite-based reagent being produced by Iv'khimprom Co. from 1992 is Ronkol. It is an efficient reducing agent for the rongalite-potash method in vat dyes printing. It provides high coloring characteristics. Application of Ronkol has made it possible to decrease reducing agents expenditure 2-3 times because of minimization of rongalite waste when printed fabrics are being dried and stored. Moreover, rate and completeness of dyes reduction as well as labour conditions now are more favorable than earlier. Ronkol's advantage over imported analogues (Rongalites C and FD) is considered to be a far less expenditure of the reducing agent. A composition of Ronkol with aforementioned Lakran is extremely efficient for coloration of viscose-staple fabrics, so its application broadens a field of the rongalite-potash method significantly. Further works to widen assortment of reducing agents for vat dyeing of cellulose materials is being carried out on the basis of the general principles recapitulated above.

The third powerful tool to affect a state and chemical activity of dyes and auxiliaries in aqueous media is mechanochemical and magnetic activations of liquid-phase systems.

Application of magnetic treatment in textile industry remains non-conventional and is rarely used in spite of phenomenal results obtained in thermal power engineering, agriculture, and medicine. However, considering complexity of textile technological processes, both nature of an object to be processed with magnetic field and a moment of its switch-on should be substantiated theoretically.

Complex investigations on magnetic activation of compound chemicals are being carried out in the INASC. Hydrotropic properties of urea, chemical activity of reactive dyes, Carbamole (based on dimethylol ethylene urea), Chromolane (chromic acetate/stearate), complex formation ability of dyes with urea and amino-formaldehyde resins, polycondensation reactions are favoured by the magnetic field treatment [11, 12].

Fig. 3 illustrates variants of magnetic activation in dyeing and finishing of fabrics. They comprise not only conventional magnetizing of water and a bath medium (variants I and II) but more efficient one of a textile material and bath solution during impregnation (III), after wringing (IV), on the stages of drying (V) and fixation (VI). Realization of these variants in accordance with complex or individual scheme is determined by technological peculiarities of a process. For all variants suggested, experimental units of equipment are developed and being tested on textile factories.

Two examples of magnetic activation tested industrially, namely magnetochemical stabilization of reducing compositions and thermomagnetic fixation of dyes and finishing agents, should be emphasized.

The first method (Fig. 4) is based on a labile change in aqueous-organic interfacial properties as a result of magnetically induced fluctuations of reagents concentrations [13]. As an efficient ingredient for sodium dithionite solutions, a new reagent called by Reagent F being a product of oxyethylation/oxypropylation of phenol is suggested. In a course of magnetic treatment of solutions it gets surface-active and hence inhibits oxygen absorption by solutions (Fig. 4a). Kinetics of sodium dithionite oxidation (Fig. 4b) shows

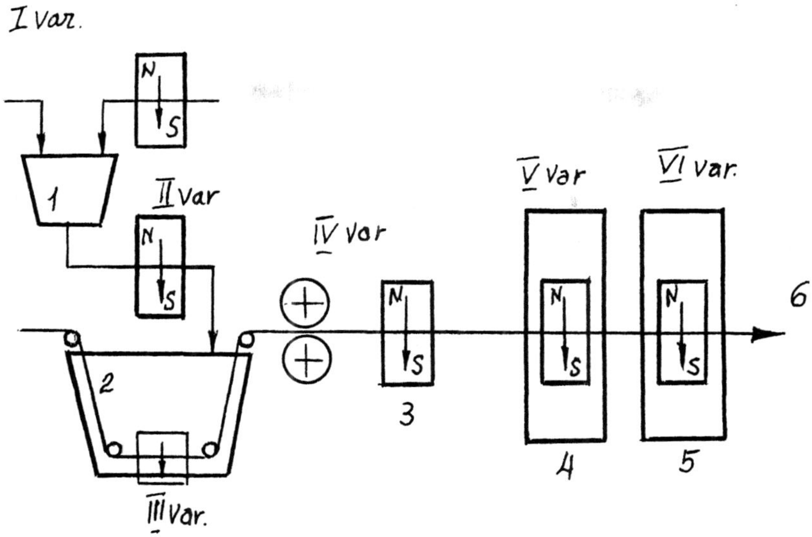

Fig 3. Magnetic activation variants for dyeing and finishing stages: 1. mixer; 2. padding; 3. wringing; 4. drying chamber; 5. thermal chamber; 6. fabric

that magnetochemical stabilization makes it possible to reduce waste of sodium dithionite down to 20% during 6 h. storage (curve 1). The waste without the stabilizer is twice as much. Already known chemical methods of stabilization, e. g. ones based on sodium sulfide application (curve 2), are less efficient than the new one being besides unfavorable medically.

The approach suggested can be realized in synthesis and application of reducing agents. In particular, its realization on a stage of magnetochemical reduction in suspension dyeing with vat dyes makes it possible to reduce sodium hydrosulfite expenditure by 25% and to raise sharply reproducibility of colors [14].

Inspiring results have been achieved also by magnetothermal treatment of reagents melts on a stage of fixation of dyes and finishing agents (Fig. 5). Magnetic activation is more efficient at low temperatures and makes it possible to get desired technological results at 140-150°C during 30-40 sec. Investigation of the finishing agents state in fabrics shows significant drop in concentration of both active methylole groups and free formaldehyde [13].

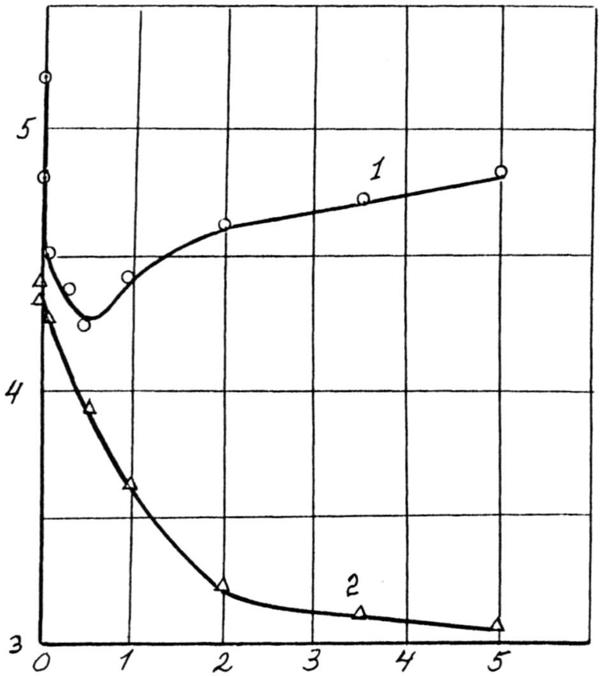

y - oxidizing rate, mol/m^2S
x - preparation P concentration, g/l

Fig 4a. Influence of magnetic treatment on inhibiting activity of Reagent F in oxidation of sodium dithionite (a) and on stability of the reducing solutions (b): 1. magnetochemical composition of sodium dithionite (35 g/l) - sodium hydroxide (20 g/l) - Reagent F (2 g/l); 2. chemical composition of sodium dithionite (35 g/l) - sodium hydroxide (20 g/l) - sodium sulfide (10 g/l); 3. conventional composition of sodium dithionite (35 g/l) - sodium hydroxide (20 g/l)

Modernization of textile equipment for IR-treatment of fabrics has been carried out on textile factories in Ivanovo and Moscow by means of installation of electromagnetic blocks. It has reduced expenditure of energy 1.5 times and one of chemicals by 20-25%. Quality of fabrics has been enhanced as well.

The modernized equipment is devoid of shortages standard IR-chambers have, namely instability of results because of anisothermic conditions and materials thermal destruction caused by overheating. Nowadays principally new types of equipment for thermomagnetic treatment by inertia-free halogenous irradiators are being developed.

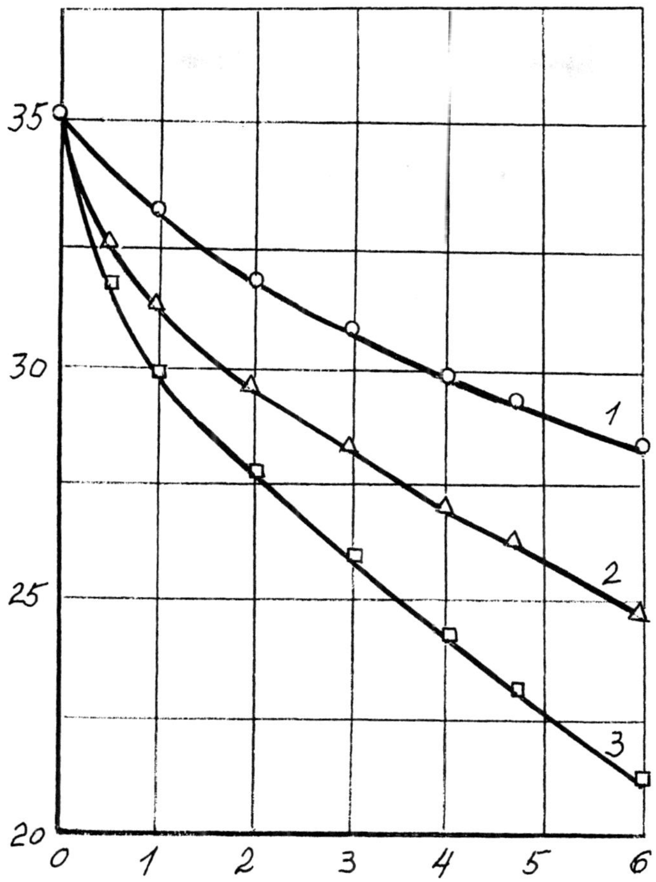

y - concentration of $Ma_2S_2O_4$,g/l
x - the time of preserving the solution, hour

Fig. 4b

Mechanochemical activation run by rotor-impuls activators developed in the INASC is not only due to mechanical rotation of liquid with velocity of (2-10)*10^3 min^{-1} but includes cavitation as well. The latter initiates sound and ultra-sound waves which generate radicals, ions, electrons, etc. Application of these activators to aqueous dyeing suspension of disperse and vat dyes makes it possible to control disperse state of dyes by introduction of special organic additives and regulation of activation parameters [16]. For example, mechanochemical treatment of a disperse dye suspension in water

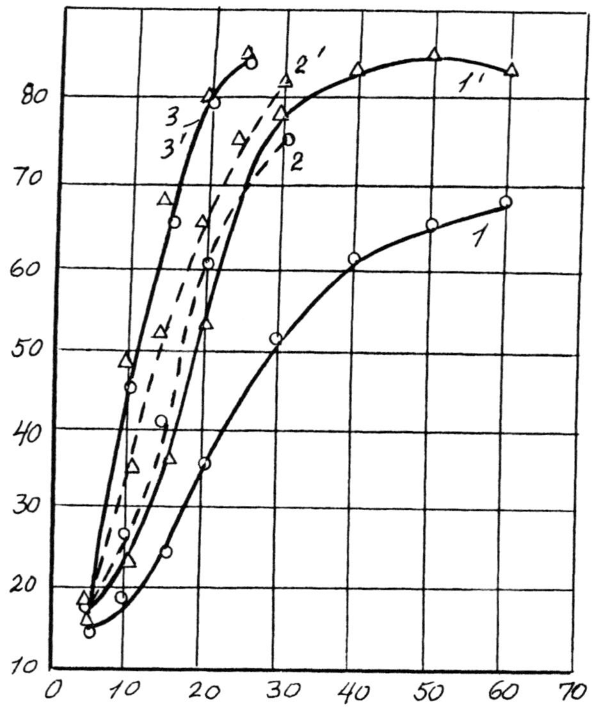

y - the degree of the preparation fixation, %
x - time(T), S

Fig 5. Fixation of a dressing based on Carbamole in IR-treatment without activation (a) and in magnetic field (b) at 140°C (1), 160°C (2) and 180°C (3)

promotes intensive size-reduction of coarse-sized particles (Table 6). Amount of fine-sized particles increases by 10-20% depending on the fractional composition of a dye ready form. A raise of the total surface value of the particles causes reactivity to increase. Therefore, dye reduction period drops by 30%. Introduction of a thickening agent into a dyeing system, e. g. sodium alginate (antimigrant), changes sharply the particles size distribution due to formation of labile associates "dye - thickening agent". Mechanochemical activation produces a fine-sized form of a dye raising its surface energy and, thus, makes a dye to associate with alginate macromolecules giving three-dimensional structures. It is supported by a sharp drop in highly fine-sized particles fraction (<10 micrometer) and a rise in 20-40 micrometer fraction (Table 6).

As a result, this method of activation being applied in dyeing of synthetic and cellulose fibers with disperse and vat dyes promotes intensification of coloring and dyes expenditure reduction by 10-30 and 5-20%, correspondingly. Moreover, dyes migration on a stage of drying of fabrics impregnated with dyeing solutions drops 1.5-3 times [17].

Table 6.

Influence of Mechanochemical Activation and Dyeing Bath Composition on the State of Vat dyes in Aqueous Suspension

Dye	Activation	Percent of dye particles with size (micrometres)				
		>10	10-16	16-40	40-100	>100
Vat Brilliant Green	-	81.4	9.3	2.5	3.7	0.4
	+	96.3	2.6	1.1	0	0
Vat Brilliant Green	-	29.5	5.3	20.3	43.1	1.8
with thickening agent	+	5.5	30.1	41.9	21.7	0.8
Vat Brilliant Blue	-	68.8	21.6	8.5	0.7	0.4
	+	90.4	6.9	1.8	0.6	0.1
Vat Brilliant Blue	-	17.1	3.7	53.8	24.7	0.7
with thickening agent	+	6.9	5.6	69.4	17.5	0.6

Note: the titles of the dyes are given in Russian variant

Activation of printing compositions based on vat and disperse dyes raises coloring parameters and fastness of colors (Table 7) [18]. However, the most important point is considered to be a possibility to control viscosity and thixotropy of high-molecular systems (Fig. 6) and consequently to raise their thickening ability [19].

Mechanochemical treatment is expedient to be applied in sizing agents preparation. For example, mechanoactivated starch sizing makes it possible a) to use compositions devoid of chemical breaking agents, b) to give up a boiling stage, c) to improve technological properties of a size [20]. It makes the new method of size production favorable ecologically and economically.

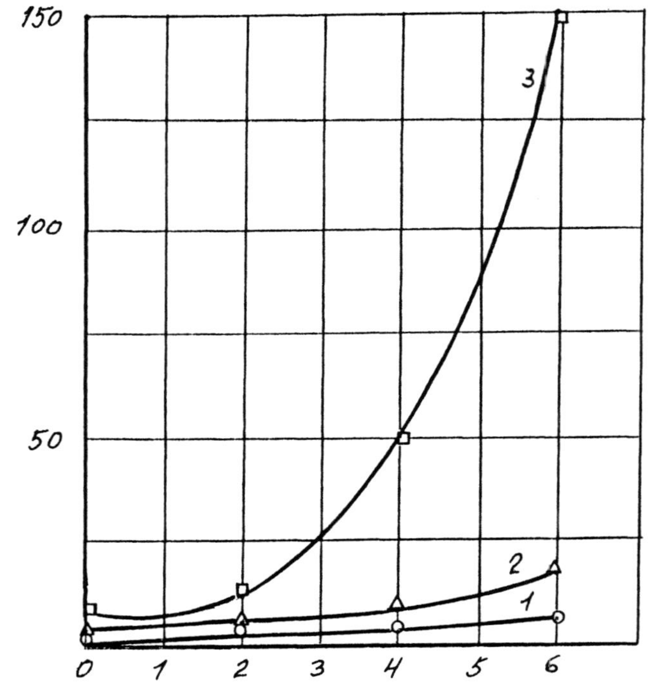

y - Dynamic viscosity (η) mPa •S
x - frequency of rotor gyration, uv/min

Fig 6. Influence of mechanochemical activation on dynamic viscosity of the thickening agent based on carboxymethyl cellulose at its content of 6% (1), 7% (2) and 8% (3)

Table 7.
Influence of Mechanochemical Activation in Cotton Fabrics Printing

Dye	Rotor speed min^{-1}	Color characteristics			Color fastness number for		
		Lightness	Saturation units	Tint, deg.	Wash	Sweat	Dry rubbing
Vat Brilliant Purple	-	38.6	73.4	183.9	5/5	5/5	3-4
	2	35.3	76.7	184.6	5/5	5/5	4-3
	4	34.8	77.2	184.0	5/5	5/5	4-3
	6	33.5	77.5	184.2	5/5	5/5	4
Diazole Ruby (on azotolized fabrics)	-	40.0	70.6	25.1	5/4	5/4	3
	2	39.6	71.3	24.7	5/4	5/4	3-4
	4	38.6	71.7	24.8	5/4	5/4	4-3

Nowadays in the INASC a new full-sized block-and-module system based on a rotor-impulse apparatus is being developed for preparation and mechanochemical activation of high-molecular compounds. Its application is appeared a) to reduce thickening agents expenditure by 10-40% in printing and sizing compositions production, b) to improve quality of fabrics and yarns, c) to reduce application of chemicals imported, etc. The complex of equipment being developed is designed for production of principally new compositions with unique properties, e. g. ones from a mixture of several thickening agents or ones from a thickening agent and a modifier which are incompatible under usual conditions. A field of compositions application is evidently not only textile industry but oil mining, chemical and food industries and so on.

References

1. A. P. Moryganov and B. N. Mel'nikov. Izv. Vyssh. *New Approaches to Development of Liquid-Phase Compositions in Textile Industry* Uchebn. Zaved. Tekhnol. Tekst. Promyshl., N 4, 57 (1991) (in Russian).

2. A. E. Khudiakov, A. V. Baranov, A. P. Moryganov and B. N. Mel'nikov *Plasticization of Cellulose-Based Textile Materials in Thermofixation of Reactive Dyes.* Izv. Vyssh. Uchebn. Zaved. Khim. i Khim. Tekhnol., *33*, 106 (1990) (in Russian).

3. A. E. Khudiakov, A. V. Baranov, A. P. Moryganov and B. N. Mel'nikov *An Efficient Mixed Plasticizer System for Reactive Dyes Thermofixation.* Am. Dyestuff Rep., *80*, 50 (1991).

4. N. A. Yakunin, A. P. Moryganov and B. N. Mel'nikov *Investigation of Non-Additivity Origin of Cellulose Triacetate Plasticization by Water-Alcohol Mixtures.* Visokomol. Soed., *33 A*, 470 (1991) (in Russian).

5. SU 1525239 (1989). *A Composition for Coloration of Mixed Textile Materials based on Cellulose Acetate - Synthetic Polymer Fibers.*(in Russian).

6. N. P. Prorokova, S. Yu. Vavilova, S. M. Prusova and Yu. A. Kalinnikov *A New Composition for Periodic Dyeing of Polyester Fibrous Materials with Disperse Dyes. Int. Conf.* "Textile Chemistry", Book of Abstr., Ivanovo, Oct. 1992, p. 20 (in Russian).

7. E. L. Kadikova, S. M. Gubina and B. N. Mel'nikov *Mechanism of Starch Polysaccharides Hydrolysis by Nucleophylic Particles Action*. Izv. Vyssh. Uchebn. Zaved. Khim. i Khim. Tekhnol., *34*, 88 (1991) (in Russian).

8. SU 1186718 (1985). *A Method of Starch-Sized Cellulose-Based Materials Desizing* (in Russian).

9. V. G. Stokozenko, S. M. Gubina and B. N. Mel'nikov *Cold Desizing Efficiency Raising*. Tekst. Promyshl., N 2, 57 (1985) (in Russian).

10. S. M. Gubina, V. G. Stokozenko and O. M. Mesnik *"Rontex" as a New Efficient Reagent for Cotton Materials Preparation*. Tekst. Promyshl., N 4, 60 (1991) (in Russian).

11. S. A. Koksharov, O. I. Konstantinov, A. P. Moryganov and B. N. Mel'nikov *Magnetic Field Effect on State of Reactive Dyes in Solution*. Zh. Priklad. Khimii., *63,* 565 (1990) (in Russian).

12. S. A. Koksharov, A. P. Moryganov, O. I. Konstantinov and B. N. Mel'nikov *Magnetic Treatment Application to Improve Textile Dyeing and Finishing. Khlopchatobumazhnaya Promyshl.,* N 6, 60 (1989) (in Russian).

13. V. N. Ivanov, S. A. Koksharov and B. N. Mel'nikov *An Efficient Method to Stabilize Reducing Solutions Based on Sodium Dithionite*. Zh. Priklad. Khimii., *64*, 743 (1991) (in Russian).

14. SU 1659554 (1991). *A Method of Textile Dyeing with Vat Dyes* (in Russian)

15. Russ. Pat., Claim N 4931094/12. *A Method of Finishing of Cellulose-Based Textile Material Being Moved* (in Russian).

16. M. V. Semenov, O. K. Smirnova and O. V. Kozlova *A New Method to Intensify Textile Vat Dyeing*. Izv. Vyssh. Uchebn. Zaved. Tekhnol. Tekst. Promyshl., N 1, 71 (1991) (in Russian).

17. SU 1719501 (1992). *A Method of Textile Dyeing* (in Russian).

18. SU 1796726 (1993). *A Method of Printing on Cellulose and Cellulose Acetate Materials* (in Russian).

19 .I.V.Nuzhdina, A.P.Moryganov and V.A. Padokhin *Mechanochemical Modification of Starch Thickeners*. Int. Conf. "Textile Chemistry", Book of Abstr. Ivanovo, Oct. 1992, p. 20 (in Russian).

20. I. M. Lipatova, I. L. Sedova and N. A. Ermolaeva *Mechanical Modification of Starch Size*. Tekst. Khimiya, N 1, 20 (1993) (in Russian).

Institute of Non-Aqueous Solutions Chemistry,
Russian Academy of Sciences, Ivanovo, Russia

APPLICATION OF STEAM TREATMENT TO PROMOTE FINISHING OF NON-CAPILLARY MATERIALS

M. N. Gerasimov

Theoretical aspects and results of experimental assessment of possible routes to promote impregnation of non-capillary textile materials are analyzed. Pre-steaming of the materials is supposed to be a useful method to intensify impregnation. Kinetic equations describing impregnation preceded by steaming have been developed. Examples of the approach suggested are being considered.

A raise in high-grade textile products output provided reduction of materials and energy expenditures is dependent to a great extent on efficiency of textile materials finishing. A great part of industrial methods of finishing includes a stage of impregnation with solutions or highly dispersed suspensions of finishing agents. Thus, quality of a finished textile material depends on uniformity and completeness of filling of its capillaries and pores with a finishing solution.

Mostly, textile materials are being treated in their coarse form, the liquid-phase processes being sizing, desizing, mercerization, boiling, etc. Because of economical and technological reasons, there exists a trend to raise output of coarse materials treatment in dyeing and finishing. However, because of poor wettability of natural and synthetic fibers (cotton, flax, polyamide, polyester), great problems arise concerning a finishing quality level desired. In practice they cause one to run the processes at lower rates and higher concentrations of solutions, to introduce surfactants and to substitute water with organic solvents. As a result, a drop in equipment productivity, a raise in materials and energy expenditure, environment deterioration in finishing shops take place.

Capillary-porous structure of fibrous materials is a sum of pores and capillaries of individual filaments which arose when these materials were being formed. Total porosity of textile materials, i. e. relation of pores volume to material volume, is independent practically of a fiber type being well within 0.37-0.47 for yarns and 0.70-0.81 for fabrics [1-4]. Such independence exists because 95% pores volume in textile materials belongs to macropores with diameter more than 10^{-7} m [5] which arose on a stage of materials formation, formation methods being general ones for all types of fibers (spinning, weaving, etc.). Total volume of micropores with diameters less than 10^{-8} m is not more than 3% of the material total pores volume [6].

In practice a textile material is characterized by different periods of impregnation in different liquids. E. g., the filled pores volume of a coarse cotton fabric after soaking with water during several hours is far less than one of a fabric impregnated with alcohol during several seconds [7]. An extent of a textile material pre-treatment has an impact on impregnation period as well. In particular, a cotton fabric boiled in alkaline solution can be impregnated far more rapidly than one without boiling, impregnation period being 10-100 times shorter than that before the treatment. Because a substitution of a bath liquid composition with another one or preliminary alkaline treatment of a material do not influence its porosity, one can infer that an impregnation process is influenced by material structural characteristics in a far less extent than by capillary and surface phenomena brought to existence by material-liquid interactions. Investigation of impregnation ability of liquids with the different contact angle values (cos θ) [8] has shown a sharp drop in an impregnation extent at cos θ<0.8. A lot of works on determination of the contact angle values for various textile materials has given inconsistent results. However, one should mark some common regularities. For example, a coarse cotton and cotton textile materials have contact angles of wetting with water far more than 60° (a fiber has 82.5^{o} [9], a compressed fabric has 60° [8]). This fact is appeared to be an evidence for strong hydrophobic properties of natural cotton fibers caused by waxes on their surface. High hydrophobic properties are inherent to a surface of bast-fibers and synthetic fibers which are used broadly in textile production. In particular, the contact angle values for fiber-water interface are 0.07 for polypropylene [10], 0.37 for polyamide [9] and 0.19 for cellulose acetate [11]. Introduction of 25% Dacron in a cotton fabric impairs sharply its capillary properties [12].

It is known [13] that there is a lot of blind pores and capillaries in textile materials. Therefore, air pinched in them prevents them from filling with liquids. Conventionally a filling process includes two stages. During the first one, a rate of capillaries filling is controlled by a pressure difference ΔP [14]:

$\Delta P = P_0 + \Delta P_c - P_0(T_l/T_0)[l_0/(l_0 - l)]$,

where P_0 is pressure over a liquid intended for impregnation;

ΔP_c is pressure in a capillary of radius r being filled with a liquid of surface tension σ and contact angle cos θ;

T_l, T_0 are temperatures of a liquid and a medium in a capillary before impregnation;

l_0, l are lengths of a capillary and its filled part.

Pressure in a capillary can be calculated as

$$\Delta P_c = 2\sigma \cos \theta / r$$

The first stage of blind pores filling is completed at $\Delta P=0$. The limiting length l_∞ of a blind pore filling at this stage can be calculated if $T_l = T_0$ as

$$l_\infty = P_o l_o / (\Delta P_c + P_o) \quad (1)$$

Analysis of eq. (1) proves that a reduction in cos θ (when wetting of an inner capillary surface decreases) causes the limiting length of a blind pore to decrease as well. This effect entails a raise in pinched air volume after the first stage is over.

Calculations [15] prove that the first stage period is not longer than 10^{-2} sec. Thus, impregnation kinetics is controlled by the second stage, i. e. filling of pores containing pinched air. The impregnation mechanisms on the first and second stages are quite different. According to the Henry law, gas solubility is proportional to pressure. So long as pressure of pinched air is higher than one over a liquid surface, concentration of air under a capillary meniscus is higher than that in a liquid outside of the capillary. It gives rise to an air stream diffusing from the capillary and an on-coming water stream. The diffusion stage of impregnation continues up to full dissolving of air and filling of capillaries with a liquid [16]. When textile materials are being treated with poorly wetting liquids, so called quasi blind capillaries develop at the first stage of impregnation. They come up because capillary forces decrease and some pores and capillaries retain pinched air despite of their openness. Percentage of the quasi blind structures in poorly wettable materials reaches 60% [17].

The aforesaid proves that period of fibrous materials impregnation with poorly wetting liquids is controlled by the diffusion stage and depends mainly on a pinched air volume remained after the first stage. This dependence is supported by eq. (2) [18]:

$$t=[\Delta l(0.25\pi D)^{0.5}/hDRT]^2, \qquad (2)$$

where $l=l_0-l_\infty$ is length of a capillary's unfilled part after the first stage is over;

D is a coefficient of gas diffusion in a liquid intended for impregnation;

h is the Henry constant;

R is the universal gas constant.

Considering eq. (2), one can reduce period of the diffusion stage by two means, namely raising D or reducing quantity of air (that is $\Delta.l$) pinched after the first stage.

Dependence of the diffusion coefficient on temperature is

$$D=kT/(6\pi\mu r), \qquad (3)$$

where k is the Boltzmann constant;

μ is dynamic viscosity;

r is a radius of particles diffusing.

Calculations based on eq.'s (2) and (3) prove it possible to reduce the diffusion stage period 7-10 times raising temperature from 20 to 100°C. Realization of this direction requires expensive steam equipment. However, in this case steam treatment takes not rarely up to 1 min.

A reduction in Δl after the first stage is connected with a limiting impregnation extent φ_∞ which can be calculated for a case of atmospheric impregnation as [14]:

$$\varphi_\infty=l_\infty/l_o=1-[P_{rem}/(P_{atm}+\Delta P_c)](T_l/T_m) \quad (4)$$

Analysis of eq. (4) proves that possible routs to raise φ_∞ are as follows: a) to raise pressure ΔP_c in capillaries; b) to reduce relation T_l/T_m before impregnation; c) to reduce pressure P_{rem} remained in pores before impregnation.

To raise capillary pressure when poorly wettable materials are being treated, it is necessary to reduce contact angle of a material surface. In practice one adds surfactants into liquid solutions [19] or substitutes water with lower alcohols [20]. A change in capillary pressure is caused by a significant change in surface tension for organic-aqueous mixtures. Factors impeding treatment of non-capillary textile materials with such mixtures are a high cost of special equipment to apply and utilize organic solvents as well as problems connected with biochemical purification of waste water containing surfactants.

The most efficient direction to raise the φ_∞ value is supposed to be a reduction in pores residual pressure before impregnation. This direction can be realized by pre-treatment of a material to evacuate air from pores. Some empirical results concerning estimation of efficiency of pre-treatment by evacuation, heating and steaming are given in [21, 22]. To characterize an extent of filling, a parameter U_{im} has been introduced. Its value ranges from 0 to 1 and can be calculated as a relation of a water quantity bound in capillaries to a limiting water quantity that a material is able to hold:

$$U_{im}=(u_b-u_{eq})/(u_b^{max}-u_{eq}),$$

where u_b is a residual moisture content after centrifugation of a material (corresponding to the impregnation conditions being investigated);

u_b^{max} is a residual moisture content after centrifugation of a material;

u_{eq} is an equilibrium moisture content before impregnation.

A method of U_{im} determination which provides the U_{im} values within confidential interval 0.02 at probability level 0.95 is described elsewhere [23].

Comparative investigation of impregnation methods under the conditions aforementioned has been carried out on samples of coarse cotton fabrics of trade marks (TM) 3217, 3006, 7071 and 115. The fabrics have zero capillarity and comprise a whole range of fiber characteristics (thickness, surface density, etc.). Dependence of an impregnation extent on impregnation parameters is the same one for all samples investigated. Let us consider it taking only one sample as an example (Fig. 1). The impregnation extent value U_{im}=1 is achieved when pre-evacuation starting from vacuum depth 80% (curve 3) or preliminary 3 sec. steaming (curve 1) have been used. To achieve the same extent without intensification, the fabric should be soaked during not less than 300 sec. (curve 2). Impregnation with pre-heating (curve 4) gives only U_{im}=0.75 even at 200°C, vacuum in pores being not more than 300 Torr.

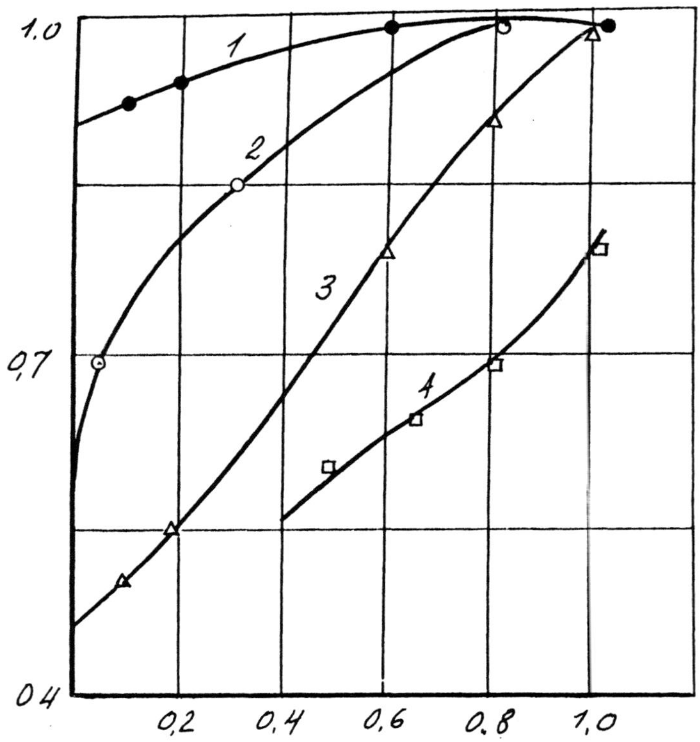

Fig.1. Influence of treatment conditions on an impregnation extent of a coarse cotton fabric (TM 3217):

1. axis OX is a steaming period before impregnation; 1 unit is equal 5 sec;
2. axis OX is an impregnation period without activation; 1 unit is equal to 300 sec;
3. axis OX is vacuum depth in a fabric before impregnation; 1 unit is equal to 100%;
4. axis OX is fabric temperature before impregnation; 1 unit is equal to 200°C;

Impregnation kinetics for the samples [24] has been calculated as a dependence of relative impregnation period t_r on relative vacuum A being developed in pores during impregnation (Fig. 2). Relative impregnation time is considered to be a relation

$t_r = t_{int}/t$,

where t, t_{int} are periods of impregnation to the same extent under conventional and intensified conditions, correspondingly. Relative vacuum is supposed to be equal to a relation of a pressure reduction through intensive pre-treatment to initial pressure in pores. One has $A=-P_{vac}/P_{atm}$ under pre-evacuation, $A=(T_l/T_m)-1$ under pre-heating and $A=(C_{air}T_l/T_{sat})-1$ under pre-steaming, where P_{vac} is vacuum in a vacuum chamber;

T_m is temperature of a material before treatment;

C_{air} is mean relative concentration of air in pores after heating during a period t^*;

T_{sat} is temperature of saturated vapors of a liquid at impregnation pressure.

For all methods of intensified impregnation, dependence t_r *vs.* A can be described with eq. (5)

$$t_r=1/(1-A)^z \quad (5)$$

Approximation of the experimental data (Fig. 2) by eq. (5) gives the z values 8.6 under evacuation and 11.0 under pre-heating or pre-steaming. The more z value in eq. (5) and, thus, the less t_r value is, the shorter impregnation period (at equal relative vacuum in pores) is expected. Therefore, in addition to evacuation of air from pores and creation of vacuum, pre-heating and pre-steaming of a fibrous material have an extra effect due to temperature enhancing. In both cases heating of a material before impregnation up to 100°C favors liquid transport through its structure because of a raise in pores diameter, viscosity reduction, etc. The extra effect of heating and steaming on impregnation is confirmed by dependence of impregnation on vacuum in pores (Fig. 3).

Technological efficiency of different physical and chemical methods to promote cotton fabrics impregnation under the conditions of alkaline treatment has been analyzed comparatively in [25]. Impregnation with aqueous, ethanol and propanol solutions at 60°C has been applied to a coarse calico (TM 274) followed by 100% wringing. Impregnation period in alkaline solution has been 15 and 30 sec. Quality of liquid-phase treatment has been estimated by impregnation coefficient being relation of a real alkali weight on a fabric to a calculated one provided absence of interaction between alkali and a fiber.

Table 1.
Coefficients of fabric impregnation

Alcali concentration in solution, g/l	Impregnation method			
	Aqueous conventional	Based on ethanol	Based on propanol	Aqueous intensified
30	1.52	1.91	1.80	1.79
60	1.42	1.58	1.49	1.59
180	1.36	1.51	1.41	1.55

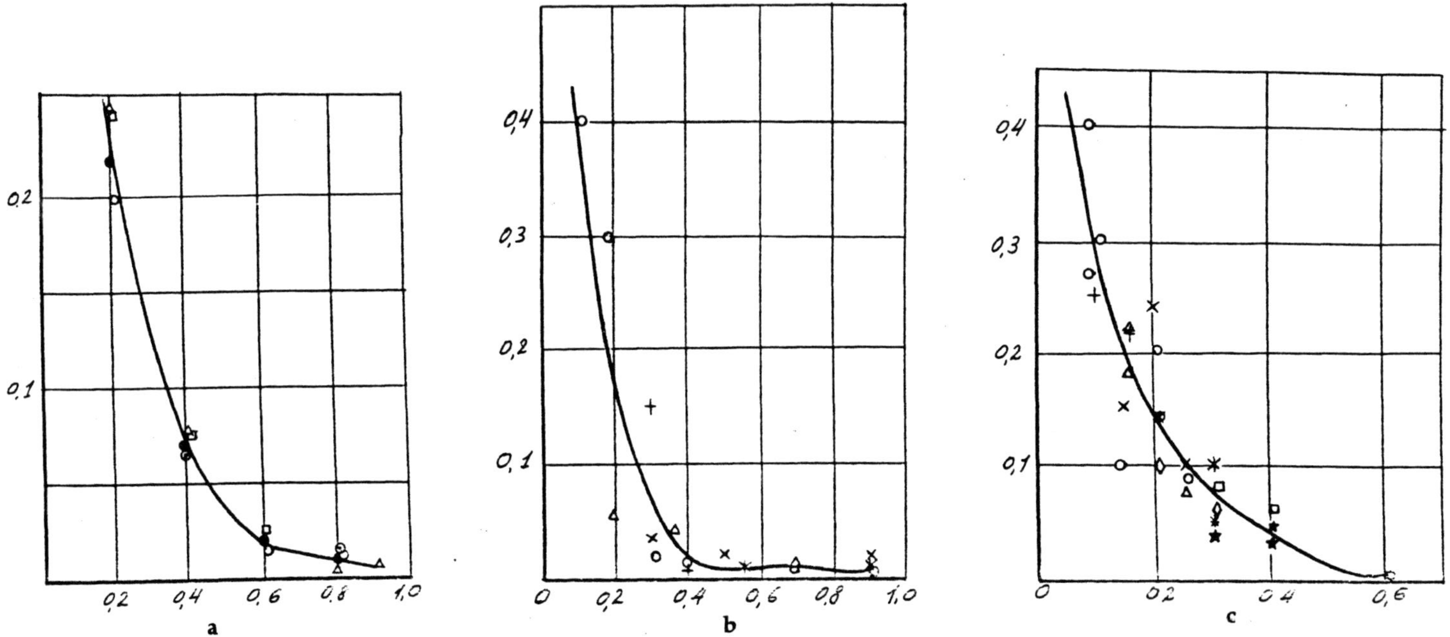

Fig.2. Influence of relative vacuum in pores (A) on relative period of fabrics impregnation with various methods of pre-treatment:

a. evacuation;

b. heating by various methods (radiation, convection, conductive);

c. steaming during various periods (t^*=0.5-5.0 sec.)

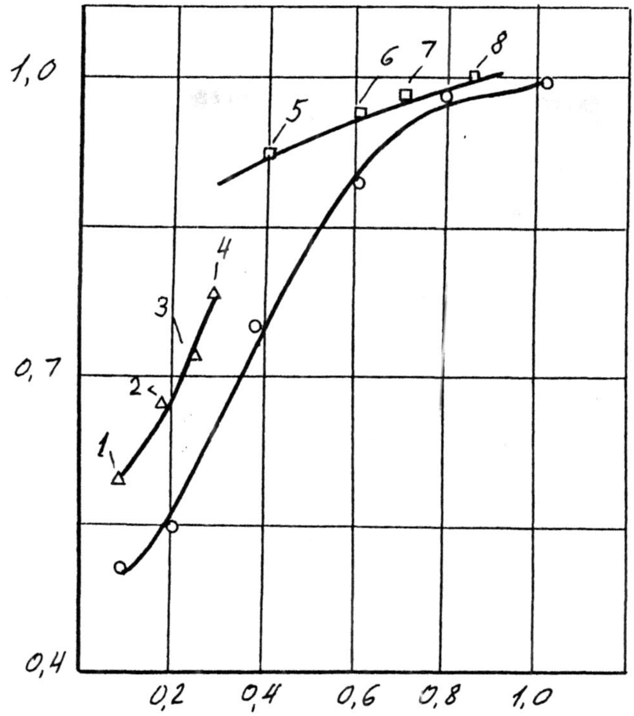

Fig. 3. Influence of relative vacuum in pores on an impregnation extent of a coarse cotton fabric (TM 3217) at 333 K during 3 sec. with various methods of pre-treatment:

a. air evacuation;

b. heating at TM 373, 403, 433 and 473 K (points from left to right);

c. steaming during period t* equal to 1, 2, 3 and 4 sec. (points from left to right);

Table 1 shows that impregnation intensification by pre-steaming is not inferior to one by alcohols being evidence of higher efficiency of the former method in a case of poorly wettable materials. Analytical investigation of wetting and heating of fibers as well as air evacuation from them during steaming [26] was based on the general theory of heat transfer in capillary-porous bodies. Experimental and theoretical efforts made it possible to derive equations for mean relative air concentration change in fibrous materials during steaming:

$$t^{*}=-\mu_{1C}\ln \quad C_{air}, \tag{6}$$

where t^* is a steaming period;

$\mu_{1,C}$ is a momentum of a kinetic curve being calculated as

$\mu_{1,C}$=$R^2/\lambda_m[a+a\text{Lu}(\text{Pn+Fe})+1/(b+1)\text{Bi}_m]$, (7)

where R is half-width of a fabric or yarn radius;

a, b are coefficients dependent on material geometry: a=1/3, b=1 for an infinite plate (fabric); a=1/8, b=1 for an infinitely long cylinder (yarn);

Lu, Pn, Fe, Bi_m are criteria of mass transfer: Lu=a_m/a_q, Pn=$-\delta(T_c-T_0)/(u_\infty. u_0)$, Fe=$\varepsilon'\delta r/c_q$, Bi=$\alpha R^2/[(\nu+1)a_m]$,

a_m, a_q are coefficients of moisture and temperature conductivity of a material;

δ is a coefficient of thermic gradient;

T_0, u_0 are temperature and moisture content in a material before steaming;

T_s is steam temperature;

u_∞ is equilibrium moisture content in a material at parameters of the vapor phase;

ε' is a criterion being equal to a relation of density of a steam flow to density of a whole moisture flow in a material;

r is steam condensation heat;

c_q is heat capacity for a material moistened;

α is a coefficient of mass interchange between steam and a material;

λ_m is a coefficient of steam conductance for a material.

For cotton fabrics [28] one has

α_m=$(1.16-0.0027B_m/2R+8.08Q)10^{-7}$ m^2sec^{-1} (8)

where B_m is surface density for a material in kg/m^2;

ρ is permeability for air in $m^3/(m^2sec)$.

Eq. (8) is true for coarse cotton fabrics when B_m is 0.15-0.40 kg/m^2, $2R$ is 0.40-0.85 mm and ρ is 0.020-0.070 $m^3/(m^2sec)$.

Coefficients a_m and δ are dependent on temperature and moisture content of a fabric [29]:

$a_{m(t)}/a_{m(298)}=(T/298)^n$

where $a_{m(298)}=f(u)$;

$n=10$ for $0.05<u<0.30$ and $n=15$ for $0.35<u<0.95$. Values of these coefficients for thin-cloth fabrics are collected in [30]. Coefficient of temperature conductivity is equal to $a_q=\lambda_q/(c_q\rho)$, where λ_q is a coefficient of heat conductivity;

c_q is specific heat capacity for a wet material: $c_q=(c_0+c_{cond}u^*)$;

c_0 is specific heat capacity for an absolutely dry material;

c_{cond} is specific heat capacity for a condensate;

u^* is mean moisture content in a material during a steaming period;

ρ is density of a dry material.

Equilibrium moisture content in a material placed in a saturated steam can be expressed as

$$u_\infty=[c_0(T_s-T_0)+(r-c_{cond}T_0)u_0]/(r-c_{cond}T_{cond})$$

Criterion of phase transition ε' can be determined from thermic properties of a material and its mean temperature. For cellulose it is estimated as 0.09-0.20 [31]. A coefficient of mass interchange α between steam and a surface of a material moving with rate V in a saturated steam is calculated from equation for a mass interchange Nusselt criterion [32]:

$$Nu_m=\alpha R/D=0.664\mathrm{Re}^{0.5}Pr_m^{0.33},$$

whereof

$$\alpha=D(0.664\mathrm{Re}^{0.5}Pr_m^{0.33})/R$$

where $\mathrm{Re}=VR/\nu$;

V is a linear rate for a material;

ν is mean kinematic viscosity for air and steam in m^2/sec;

$Pr_m=\nu/D$;

D is a coefficient of mutual diffusion of air and steam in accordance with eq. $D=D_0(T/273)^n$, where $D_0=0.216\ 10^{-5}$ m^2/sec, $n=1.8$ and T is temperature of their mixture within 273-1493 K.

The equations suggested and the method to calculate air evacuation kinetics during steaming of a textile material have been tested empirically. Fig. 4 shows experimental

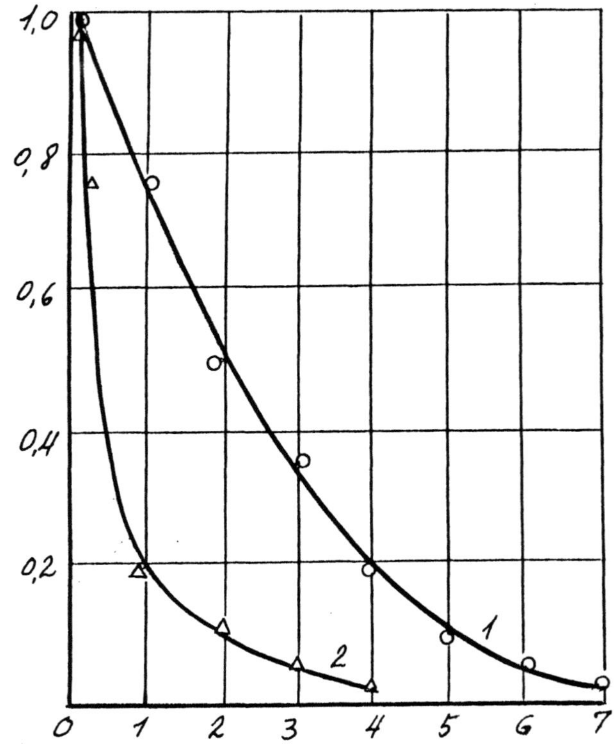

y - Relative air concentration, θ air
x - Duration of steaming, S

Fig. 4. Kinetics of air evacuation from pores in treatment with saturated steam for serge, TM 3217 (1) and repp, TM 871 (2)

data (points) for relative concentration of air in pores *vs.* a period of steaming and curves C_{air} *vs.* t^* calculated by means of eq.'s (6) and (7). Results presented have been used to work out new technologies of textile finishing as well as new equipment for pre-steaming by impregnation. Originality of this technical approach is supported with nine patents.

Fig. 5 shows schemes of equipment for liquid-phase treatment of textile materials proceeded by steaming. Fig. 5a gives details of construction of a unit for intensified impregnation of non-capillary fabrics in bleaching and dyeing as well as imparting of special properties to fabrics (crease-resistance, water-resistance, fire-proofing, etc.). Peculiarities of textile materials treatment with concentrated alkaline solutions (mercerization) are taken into account by another construction (Fig. 5b). Further, Fig. 5c

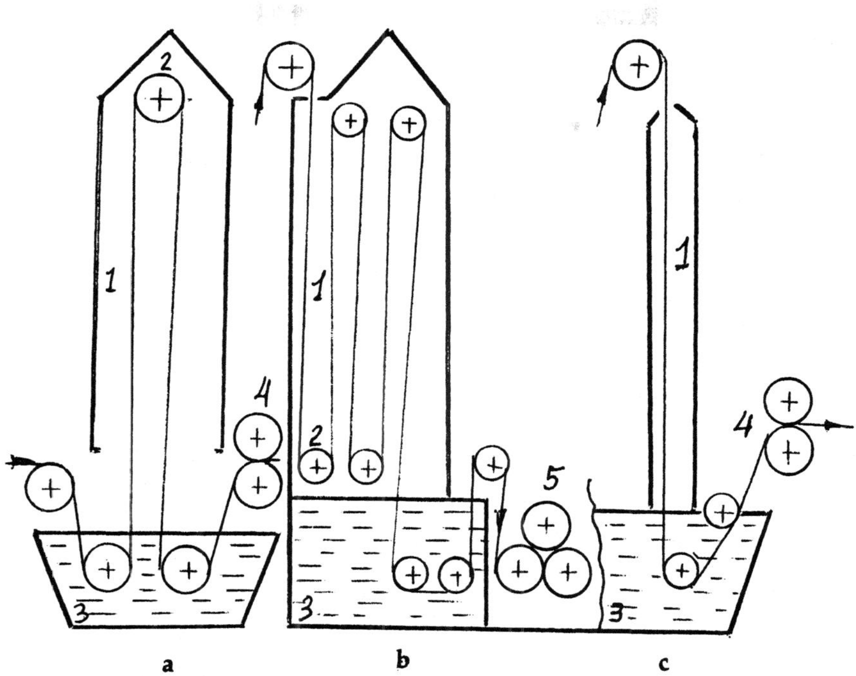

Fig. 5. Equipment for textile materials impregnation with pre-steaming (1. steam chamber, 2. roller, 3. solution for impregnation, 4. bath, 5. wringer rollers, 6. mercerization unit)

shows a unit for liquid-phase treatment of yarns with pre-steaming during sizing or dyeing.

Experimental samples of the equipment have been tested industrially. A line for vat dyeing (LKK-140-1, textile factories in Ivanovo, RF) has been modernized through replacement of IR-heating unit (KNT-140) by a pre-steaming chamber (Fig. 5a) mounted over the first dyeing bath. Comparative testing of dyeing of a coarse repp (TM 871) intensified by pre-heating and pre-steaming before impregnation have proved superiority of the method worked out in our laboratory. Pre-steaming of fabrics makes it possible to raise a dye yield into a fiber under the conditions of high-rate treatment or to raise 1.5-2.0 times a rate of treatment provided dyeing intensity already achieved (Table 2). Dyeing fastness corresponds to standards for a dyes class applied. Substitution of the

KNT-140 chamber with a pre-steaming one makes it possible not only to improve dyeing but to reduce twice energy expenditure for dyeing.

Table 2.

Influence of pre-steaming on relative dye yield into fabric

Treatment rate on LKK-140-1, m/min	Pre-treating in KNT-140	Pre-steaming
40	1	1.24
60	0.91	1.11
80	0.86	1.03

A new dyeing line LKS-140-12 (Ivtekhmash Co., Ivanovo) includes a pre-steaming chamber. Industrial testings of the line in dyeing with sulphur dyes have shown (Table 3) that pre-steaming of coarse fabrics makes it possible to raise by 20% a quantity of a dyeing solution applied to a fabric and thus to raise intensity of dyeing. Because of deep penetration of a dyeing solution, a fixation extent increases sharply and so does physicomechanical fastness of dyeing. Pre-steaming effect in dyeing with direct and reactive dyes have been tested as well. Colorings of light and middle hues are determined to be even and pure. Fastness to physicomechanical factors corresponds to the classes of highly fast dyeing for reactive dyes and normal dyeing for direct ones. A fixation extent in both cases is within 72-78%. Fig. 6 shows that pre-steaming (curves 1, 2) of non-capillary cotton fabrics before their impregnation with a dyeing solution causes a dye yield to increase 2.2-2.4 times in comparison with a conventional method (curves 3, 4).

A modernized line LBK-140 with a pre-steaming chamber mounted over the first technological bath has been tested industrially. Short pre-steaming (2-3 sec.) of a coarse fabric makes it possible to raise quality of finishing, especially capillarity and whiteness, and reduce surface impurities content with a simultaneous increase in output by 40% (Table 4).

A new method of a warp yarn sizing has been tested on two textile factories in Ivanovo. In accordance with the method, a steam treatment of a fabric during 1.5-3.0 sec. is followed by impregnation with a sizing solution (Fig. 5c). Disruption strength of a yarn treated with a steam is by 5-7%, disruption elongation is by 6-10%, many-fold

Table 3.

Efficiency of pre-steaming in coarse fabric dyeing

Fabric trade mark	Treatment rate on LKS-140, m/min	Pre-steaming period, sec	Weight increase after impreg-nation, %	Relative dye content, %	Dye fixation extent, %	Dyeing fastness to physico-chemical factors, points		
						Soap and soda, 40^{0}C	Sweat	Dry rubbing
Satin TM 546	60	-	63	100	70	4/4/4	4/4/4/	3
	60	3	89	125	82	4/5/5	4/5/5	4-3
	80	-	57	85	68	4/4/4	4/3/4	3
	80	2.2	89	120	80	4/5/5	4/4/5	4-3
Satin, TM 3080	50	-	66	100	74	4/4/4	4/4/4	3
	50	3.6	87	122	81	4/5/5	4/5/5	4-3
	60	-	62	88	71	4/3/3	4/3/4	3-2
	60	3	83	115	79	4/5/5	4/5/5	3

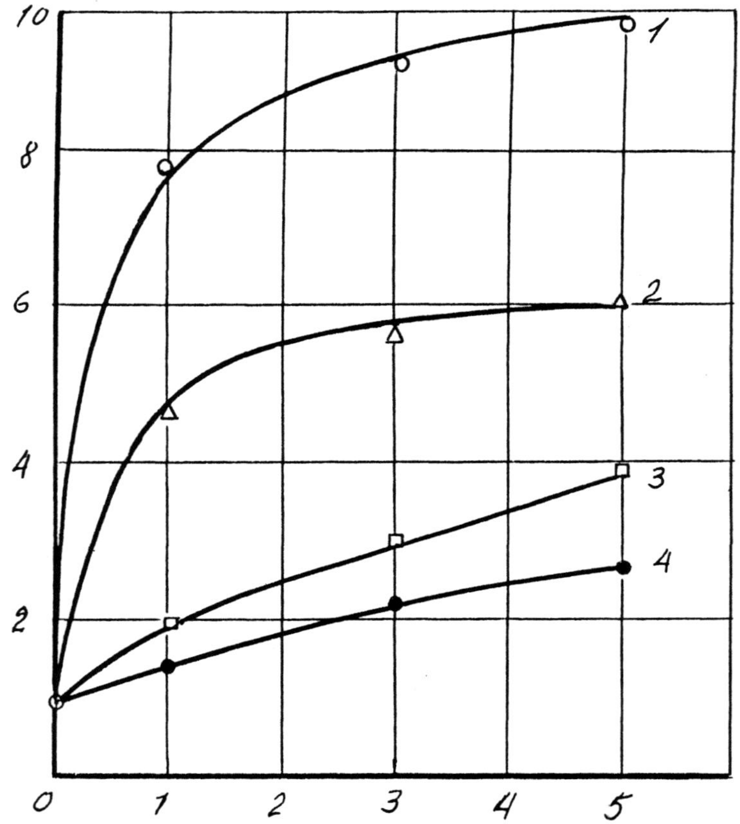

y - Dye yield, g/kg
x - Tixation time, min

Fig. 6. Kinetics of dye fixation on a cotton fabric (with surface density 356 g/m^2) for Ostazine golden-yellow (1, 3) and Direct crimson light-fast (2,4) with pre-steaming (1, 2) and without it (3, 4)

stretching resistance is by 20-26% higher than the corresponding values for a yarn sized conventionally. Such physicochemical improvement of a yarn is due to a deep penetration of a sizing solution into its activated structure. The method has been applied on 20 sizing units during a long period of time. It has been stated that breaking of the yarn on weaving looms decreases by 25-45%.

Table 4.
Efficiency of pre-steaming in coarse fabric preparation

Fabric trade mark	Impregnation method	Treatment rate, m/min	Physicomechanical properties of a fabric					
			Capillarity, mm/hour		Fastness of a band		Whiteness, %	Impurities, m^{-2}
			Warp	Weft	Warp	Weft		
Calico TM 23	Conventional	50	40	40	280	182	72.0	500-1300
	pre-steaming	70	80	70	340	215	75.6	0-800
Calico TM 20	Conventional	50	40	50	333	218	72.5	500-1800
	pre-steaming	70	88	73	370	220	78.6	0-400

High technological and economical effects due to application of pre-steaming in impregnation intensification have been gained in one-bath mercerization - boiling - vat dyeing (Table 5), in linear dimensions stabilization and crease-resistance finishing (Table 6), in water-resistance finishing (Table 7), in fire-proofing (Fig. 7), in mercerization (Table 8 [33]), in azeotropic coloration [34] of cotton, flax, viscose and polyester fibers and mixtures thereof.

Table 5.
Efficiency of pre-steaming in vat dyeing

Dye	Pre-steaming	Dyeing extent, %	Fabric quality				
			Rubbing fastness		Barite number, %	Capillarity, mm/hour	Specific viscosity
			Dry	Wet			
Light-green GD	+	43.4	5-4	3	125	100	1.17
	–	73.5	5	3	131	115	1.34
Golden-yellow GD	+	47.1	4	3	125	100	2.20
	–	67.9	4	3	127	128	1.31

Table 6.

Efficiency of pre-steaming followed by impregnation in shrink-resistant and crease-resistant finishing

Pre-steaming	Carbamol, g/l	Solution content on fabrics, %	Resin on fabrics (%) after		Formaldehyde, %		Linear dimensions change		Angle of opening, deg.	Disruptive strength ref. to a non-treated fabric		Rubbing fastness cycles,
			Finish.	Wash.	Total	Bound	Warp	Weft		Warp	Weft	
Coarse cotton fabric, TM 3355												
+	180	67	5.50	3.47	0.88	0.31	2.8	3.3	177	82.0	73.8	1992
–	180	41	3.87	3.08	0.87	0.29	3.0	1.1	173	82.7	79.9	1644
+	180	73	5.97	4.28	1.38	0.41	2.2	0	215	79.5	75.4	1115
Viscose dacron fabric, TM 82125												
–	220	84	10.20	8.00	2.13	1.69	2.6	-	202	-	-	2832
+	150	92	8.20	6.00	1.74	1.08	2.5	-	207	-	-	2902
–	150	47	4.40	3.30	0.88	0.37	3.0	-	200	-	-	3070

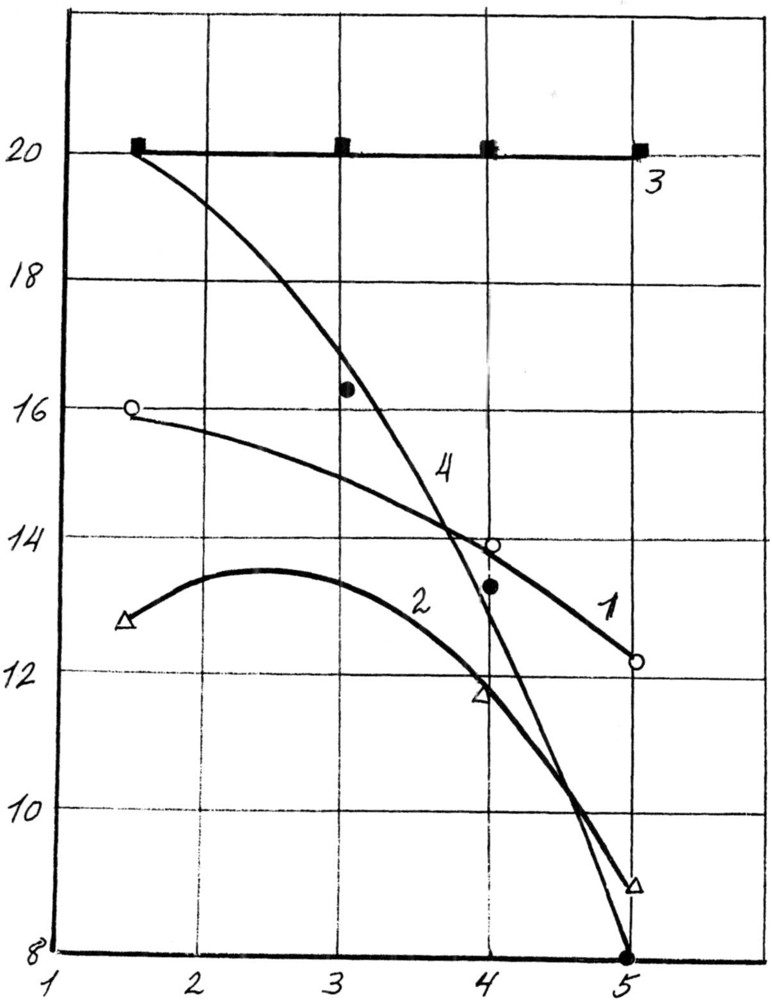

x - Impregnation time, S
y - The height of fabric cooling, cm.

Fig. 7. Influence of pre-steaming on fire-proofing properties of a fabric, TM 30, after finishing without steaming (1, 3), after finishing with steaming (2, 4) and after washing (3, 4)

Table 7.

Efficiency of pre-steaming in water-resistant finishing

Components of impregnation bath, g/l	Impregnation conditions	Treatment rate, m/min	Physicomechanical parameters				Relative amount of dressing applied to a fabric after	
			Disruption strength, N		Water resistance (Pa) after			
			Warp	Weft	Finish.	Wash.	Finish.	Wash.
Alamin[a] (60), acetic acid (6), ammonium chloride (3)	Without pre-steaming	56	565	610	202	89	4.4	1.7
	With pre-steaming	56	568	642	235	225	4.6	2.4
State standard	Not less than	-	461	530	200	140	4-5	-

Table 8

Impregnation method	Merserization parameters					Finished fabric properties (Satin TM 535)						
	Treatment rate	Conditions		Bleaching		Pulling		Barite number, %		Whiteness, %	Capillarity, mm/hour	
		NaOH, g/l	t^0C	NaOH, g/l	t^0C	Mercerized	Bleached	Mercerized	Bleached		Warp	Weft
4 sec. steam	70	200	38	62	78	1.5	1.9	136	1215	86.7	120	101
No steam	50	200	38	62	78	1.1	12.8	126	121	85.4	116	105
4 sec. Steam	70	194	34	36	66	0.5	0.8	147	135	86.2	121	145
No steam	40	195	34	36	66	0.7	1.2	142	131	85.2	119	130

Application of pre-steaming on more than 140 chambers in various processes of non-capillary textile materials treatment has shown that chemicals and energy expenditures decrease by 10-15 and 20-60%, correspondingly, productivity being higher by 20-60%.

References

1. *D. P. Litevchuk* Determination of Apparent Porosity and Specific Moisture Content in Fabrics. Izv. Vyssh. Uchebn. Zaved. Tekhnol. Tekst. Promyshl., N 2, 44 (1980) (in Russian).

2. *N. S. Poroshin* Water Absorption on Cotton Fabrics. Tekst. Promyshl. N 8, 46 (1960) (in Russian).

3. *A. A. Martynova and L. A. Chernikova* Laboratory Manual on Structure and Design of Fabrics. Moscow, 1976, p. 38 (in Russian).

4. *A. A. Martynova and E. A. Yuhina* Computation and Design of Fabrics Porosity. Tekst. Promyshl., N 2, 56 (1984) (in Russian).

5. *S. S. Vojutski and L. E. Peregudova* Zh. Priklad. Khimii., *24,* 928 (1951) (in Russian).

6. *S. P. Papkov and E. Z. Fainberg* Interaction of Cellulose and Cellulose Materials with Water. Moscow, Khimiya, 1976, p. 232 (in Russian).

7. *A. N. Braslavski, V. V. Darvin, E. S. Raskin and S. N. Ivanova* On Mechanism of Fibrous-Porous Materials Impregnation under the Conditions of Blind Pore Formation and Air Pinching. Zh. Priklad. Khimii., *43,* 1803 (1970) (in Russian).

8. *R. M. Panich and S. S. Vojutski* Rep. Inst. of Fine Chem. Technol., Moscow, Sovetskaya Nauka, Issue 3, 42 (1952) in Russian).

9. *T. H. Grindstuff and H. T. Petterson* Text. Res. J., *39*, 958 (1969); *45*, 760 (1975).

10. *B. Miller and R. Young* Text. Res. J., *45*, 359 (1975)

11. *C. Hornruff and R. Gartner* Milliand and Textillberichte. *39*, 681, 799, 935 (1970).

12. *N. G. Abbasova* Hygienic Properties of Mixed Fabrics Made of Cotton and Polyester. Rep. Polytech. Inst., Tashkent, Issue 222, 65 (1977) (in Russian).

13. *S. S. Vojutski* Physicochemical Foundations of Impregnation of Fibrous Materials with Polymer Dispersions. Leningrad, Khimiya, 1969, p. 336 (in Russian).

14. *F. Yu. Telegin and B. N. Mel'nikov* Mathematical Description of Textile Materials Impregnation. Izv. Vyssh. Uchebn. Zaved. Khim. i Khim. Tekhnol., *28*, 101 (1985) (in Russian).

15. F. L. Alter-Pesotski *Physical Methods of Fibers Finishing Intensification.* Zh. D. I. Mendeleev Khim. Obshch., *26*, 433 (1981) (in Russian).

16. B. V. Deriagin and M. A. Altschuler Rep. USSR Acad., *152*, 911 (1963) (in Russian).

17. C. R. Stewart, A. Libinski and R. A. Blenkarn *J. Petrol Technol.*,N 4, 383 (1961).

18. G. A. Akselrud and M. A. Altschuler *Introduction in Capillary Chemical Technology.* Moscow, Khimiya, 1983, p. 264 (in Russian).

19. V. A. Braslavski *Capillary Processes in Textile Materials.* Moscow, Legpromizdat, 1987, p. 112 (in Russian).

20. B. N. Mel'nikov and I. B. Blinicheva *Theoretical Foundations of Fibrous Materials Dyeing.* Moscow, Legkaya Industriya, 1978, p. 304 (in Russian).

21. M. N. Gerasimov, V. V. Kozlov and Yu. B. Volzhankin *Efficiency of Various Methods of Textile Impregnation.* In: Optimization of Dyeing and Finishing of Cotton Fabrics. Moscow, Cent. Res. Inst. for Inform. in Light Ind., 1986, p. 26 (in Russian).

22. SU 340458 (1972).

23. M. N. Gerasimov and V. V. Kozlov *On the Problem of Assessment of Textile Impregnation Quality.* In: Problems of New Technologies of Cotton Fabrics Finishing. Moscow, Cent. Res. Inst. of Inform. in Light Ind., 1979, p. 57 (in Russian).

24. M. N. Gerasimov *Empirical Investigation of Intensive Methods of Non-Capillary Cotton Fabrics Impregnation.* Izv. Vyssh. Uchebn. Zaved. Tekhnol. Tekst. Promyshl., N 3, 57 (1988) (in Russian).

25. V. N. Galashina, S. M. Gubina and B. N. *Mel'nikov Intensification of Textile Impregnation with Alkaline Solutions.* Izv. Vyssh. Uchebn. Zaved. Tekhnol. Tekst. Promyshl., N 5, 47 (1984) (in Russian).

26. F. Yu. Telegin, V. V. Kozlov, B. N. Mel'nikov, V. N. Blinichev and M. N. Gerasimov *Mathematical Description of Textile Materials Steam Treatment.* Izv. Vyssh. Uchebn. Zaved. Tekhnol. Tekst. Promyshl., N 6, 50 (1985) (in Russian).

27. M. N. Gerasimov *Intensification of Textile Finishing by Steam Treatment. Abstract of Thesis*, Inst. Text. Light Ind., Leningrad, 1991.

28. V. V. Kozlov, M. N. Gerasimov and F. Yu. *Telegin Empirical Determination of Mass Conductivity Coefficient for Fabrics in Steam Medium.* In: Modern Technique and Technologies of Cotton Processing and Prospects of Industry Development in 12th Piatiletke. Book of Abstracts, Ivanovo, 162 (1984) (in Russian).

29. V. D. V'jushin, M. N. Gerasimov, V. A. Gusev, Yu. G. Ershov and V. P. Kapustin *Moist Characteristics of Fabrics. Izv. Vyssh. Uchebn.* Zaved. Tekhnol. Tekst. Promyshl., N 2, 115 (1974) (in Russian).

30. E. A. Ganin, S. D. Korneev, I. P. Kornjukhin and V. I. Shcherbakov *Heat Utilizing Equipment in Textile Industry.* Moscow, Legpromizdat, 1989, p. 392 (in Russian).

31. A. V. Lykov and Yu. A. Mikhailov *Theory of Heat and Mass Transfer.* Moscow, Leningrad, Gos'energoizdat, 1963, p. 536.

32. S. P. Rudobashta *Mass Transfer in Systems with Solid Phase.* Moscow, Khimiya, 1980, p. 248 (in Russian).

33. V. A. Denisova, N. A. Zabavina, A. A. Afanasjeva and S. S. Demidov *Mercerization Intensification by Preliminary Steam Treatment.* In: Optimization of Cotton Fabrics Finishing. Moscow, Cent. Res. Inst. of Inform. in Light Ind., 1985, p. 52 (in Russian).

34. O. K. Smirnova, A. P. Moryganov, I. B. Blinicheva, B. N. Mel'nikov, V. V. Kozlov, and M. N. Gerasimov *Fabrics Dyeing Intensification.* Tekst. Promyshl., N 1, 53 (1983) (in Russian).

35. I. B. Blinicheva *Influence of Steam Treatment of Cellulose Acetate and Synthetic Fabrics on their Coloration and Electrophysical Properties.* In: Optimization of Cotton Fabrics Finishing. Moscow, Cent. Res. Inst. of Inform. in Light Ind., 1985, p. 79 (in Russian).

State Textile Academy, Ivanovo, Russia

POSSIBILITIES AND PROBLEMS OF PLASMA TREATMENTOF FABRICS AND POLYMER MATERIALS

A.I. Maximov, B.L. Gorberg, V.A. Titov

Possibilities of polymer surface modification by treatment with low-temperature plasma are discussed in this paper. Physico-chemical and applied aspects of plasma treatment are considered. Some advantages and problems concerning a combination of plasma and conventional chemical processing for fabric finishing are outlined. Plasmachemical equipment and processes being applied in textile industry in Russia are described. It is noted that a further progress in the industrial plasma processing of polymers requires the investigation of chemically active plasma as a self-consistent, self-organizing system which includes not only the initial gas along with the products of plasma-induced reactions but also a modified polymer surface layer.

Development and application of any new technology is aimed at the solution of different environmental and economical problems. A lot of these problems can be solved using a non-equilibrium low-temperature plasma. Plasma processes are dry ones and do not require water and aqueous or non-aqueous solutions.

Principal properties of non-equilibrium plasma

Plasma is a partially ionized gas. Ionization degree of the plasmas applied for materials treatment is about 10^{-6} only, but it is enough to change gas phase properties greatly. The existence of gas discharge plasma is provided by the external source energy. The energy from electromagnetic field is acquired mainly by the plasma electrons. The collisions of energetic electrons with gas molecules lead to ions generation and these ionization processes maintain the required plasma conductivity. Under stationary conditions there is a balance of rates of electron generation and loss processes. Since ionization energy of a molecule exceeds the energy of any molecule excited state, plasma electron collisions with gas molecules are accompanied by the formation of excited species in different states including radiating ones and they are accompanied by the dissociation processes as well. As a result the charged particles, neutral active species i.e. atoms, radicals, excited species and UV- photons reach the surfaces being in contact with plasma. This is the reason of high chemical activity of plasma. At the same time the mean energies of electrons and heavy particles are strongly different owing to low efficiency of energy transfer from electrons to molecules. Plasma electron mean energy corresponds to the effective temperature of about 50000-100000 K while the temperature of the neutral gas and materials being treated is as a rule not far from the room temperature. This is a low-temperature plasma or the so called cold plasma. Such a combination of low temperature of a neutral gas and high energy of the electrons makes non-equilibrium plasma an ideal instrument for the treatment of the non-thermostable materials such as polymer films, fabrics, fibres etc.

Plasmachemical processes have some features similar to those of radiation chemistry and photochemistry. This is connected first of all with the action of charged species and UV-photons on the materials under treatment. However, there is an important difference between plasmachemical processes and radiation or photochemical processes. Chemically reacting plasma is not an active particles source controlled by an operator. It is a self-organising system and its physical properties determining chemical activity of plasma depend in its turn on the plasma-induced chemical conversions.

Effects of plasma-polymer interaction

Experimental data show plasma-polymer interaction to be a multistage and multichannel process. This fact is caused by the variety of active species generated in a plasma. At the same time, the depth of all these active species action in polymers does not exceed several micrometers (Table 1).

Table 1.

Active species of cold plasma	Energy, eV	Mean depth of active species action
Atoms, radicals, excited molecules	$kT \approx 0.01\text{-}0.1$ $E \approx 5\text{-}10$	$> 0.1 \mu m$
Ions, electrons	1 - 100	Up to 10 monolayers
UV - photons	5 - 20	Up to 10 μm

Note: kT and E are translation energy and internal energy of active particles respectively

Therefore, any applications of such a plasma treatment can be connected with the change of the materials surface properties only. All possible types of treatment processes are the following: (1) removal of the polymer surface layer with the required thickness i.e. plasma etching; (2) deposition (growing) of the film with required chemical composition and properties on the polymer surface, for example by plasma polymerization or metallization; (3) changing a composition of functional groups and a polymer structure in a thin surface layer in order to obtain desired surface characteristics. All that can be considered to be a modification of surface properties.

It is evident that any applied effect of plasma treatment is a result of many physical and chemical transformations in a polymer surface layer. However, one can not always find a simple correlation between physical or/and chemical transformations and applied effects of plasma modification (Table 2).

Table 2.

Effects of plasma action on polymers.

Chemical and physical transformations	Applied effects
1. Chemical changes in a surface layer	Change in wettability, capillarity, total moisture absorption.
1A. Formation of double bonds and new functional groups, for example -OH, -C=O, -COOH, -CN, =NH, $-NH_2$ groups; polymer chain crosslinking	Increase in water repellence and oil resistance. Increase in fastness of coloring, decrease of dyeing time, increase in printability. Modification of anti-resorbtion and anti-soiling properties. Improvement of a surface adhesion for the deposition of coatings and for the production of composites. Change in surface layer solubility in organic and inorganic solvents. Fabric desizing and cleaning.
1B. Destruction processes: - scissions of polymer macromolecules, - destruction of functional groups; - formation of gaseous etching products	
1C. Other transformations: - change in effective polymerization degree and mean value of molecular mass; - change in oxidation degree of separate atoms in macromolecules, for example oxygen atoms; change in oxidative-reductive properties of the surface.	
2. Changes in a surface layer structure : - change of crystallinity degree;	Increase of polymer fastness to weathering.

- change of phase transition temperature; - plasma initiation of phase transitions; - formation of micro-domain structures; - increase of surface effective area; - change of surface porosity and permeability, - change of diffusion coefficient values for gases and liquids in polymers.	Prevention of a plasticizer migration from polymer volume to surface. Modification of properties of ion-exchange resins and ion-selective membranes. Disinfecting a polymer surface. Imparting biocompatibility to polymers.
3. Changes of physical properties of surface layer: - change of surface energy; - change of surface friction coefficients; - changes of mechanical properties; -change in absorption spectra in IR- , UV- and visible range; - change of surface conductivity, relative dielectric constant (ε) and tgδ values; - accumulation of surface charge	Increase of polymer strength, elasticity, fastness to rubbing. Decrease of wool shrinkage and felting. Increase of crease-resistance of fabrics. The change in colour and glossiness of a polymer surface. Surface smoothing and decrease of surface friction. Imparting the required surface electric properties including antistatic ones.

The information about the use of different gases for plasma treatment of polymer materials is given in Table 3 [1 - 50]. These data show the wide change of gas nature and properties influences slightly the final treatment results. Similar desired effects can be obtained in the plasmas of noble gases and in the plasmas containing hydrocarbons, oxygen or halogens as well. In our opinion it can be explained in the following way. The main part of the applied effects is caused first of all by the changing of polymer surface energy resulting from plasmachemical etching. The surface energy is an integrated characteristics depending upon the chemical composition and structure of a superficial layer. They are modified during plasma etching. Evidently the same values of the surface energy can be attained by the different combinations of chemical and morphological transformations at the polymer surface. Moreover, it has to be noted that gaseous etching products change the gas phase composition and mask essentially the initial peculiarities of the plasma. Undoubtedly, the influence of the gas phase conversions will be more important if the treatment goal is plasma synthesis of layers with required composition and properties but not simply plasma etching.

Table3.

Gas	Results of treatment	Material under treatment
The inert gases: He, Ne, Ar	Increase in wettability	PA, PE, PET, wool, cotton
	Decrease of polymer permeability to gas	PE-, PP-, PS-films
	Prevention of a plasticizer migration from polymer volume	PVC
	Increase of tensile strength	PE-, PP-, PS-films
	Increase in bonding properties	Polyester
	Elimination of surface glossiness	PVC
	Imparting the antistatic properties	PET
Air, oxygen	Increase in wettability	Cotton fabric, wool, PET-fibres, polyester
	Decrease in shrinkage and felting, increase in fastness to rubbing	Wool fabric
	Increase in wettability and intensification of an impregnation process	Grey cotton fabric
	Modification of adhesion properties for vacuum metal layer deposition and for a gluing of polymer to metals	PE, PP, POM, PET, PTFE, polyimide, polycarbonate
	Decrease of permeability to gas	PE-, PP-, PS-films
	Increase of tensile strength	PE-, PP-, PS-films
	Elimination of surface glossiness	PVC
	Improvement of antistatic properties	PET-fabric
Silanes	Decrease of permeability to gas	PE-, PP-, PS-films

	Increase of tensile strength	PE-, PP-, PS-films
Hydrocarbons: methane, ethane, butane etc.	Decrease of permeability to gas	PE-, PP-, PS-films
	Increase of tensile strength	PE-, PP-, PS-films
	Prevention of plasticizer migration from polymer volume to surface	PVC
	Improvement of antistatic properties	PET
Oxygen containing gases: CO_2+Ar	Changes of surface morphology	Carbon fabric
CO_2	Elimination of surface glossiness	PVC
	Increase in wettability	PET-fabric, wool and cotton fabrics
NO, NO_2, CO	Elimination of surface glossiness	PVC
CO, H_2O	Increase of adhesion	PET-film
N_2	Increase of adhesion	Polyester films
	Prevention of plasticizer migration from polymer volume to surface	PVC
	Decrease of permeability to gas	PE-, PP-, PS-film
	Increase of tensile strength	PE-, PP-, PS-film
	Increase in bonding properties	Polyester
	Increase in wettability	Wool, cotton and PET fabrics
	Elimination of surface glossiness	PVC
Nitrogen containing gases and vapours: Amines	Increase of off-axis strength of fibres	Aromatic polyamide
	Increase of wettabulity, adhesion and antistatic properties	
Alkylamines	Surface activation of fibres, filaments and fabrics for the production of composites with epoxy- and triazine resins	PTFE Aromatic polyamide
NH_3	Increase in wettability	Aromatic polyamide, polyester, wool, grey cotton fabric
	Prevention of a plasticizer migration from polymer volume to surface	PVC

	Increase of polymer adhesion	POM, polycarbonate, poly(acrylate), poly(phenylen oxide), PTFE
	Increase in wettability	PMMA
H_2	Increase in wettability	Wool, cotton and PET-fabrics
	Elimination of surface glossiness	PVC
	Surface smoothing	PVC
$Ar+H_2$	Increase of adhesion	Carbon fabric
Halogen containing gases:		
Cl_2	Elimination of surface glossiness	PVC
CF_4, $CF_4 + O_2$	Change of plastic surface wettability	PE, POM, PP, PTFE, PS, PMMA, PI
NF_3, BF_3, SiF_4	Surface activation before wet treatment	Wool, cotton, silk
CF_4, CHF_3, C_2F_4, $CHClF_2$, CF_3Cl, CF_3Br	Surface activation and change of surface wettability	PE, PS, PP, PTFE, PET
CF_4, CHF_3	Improvement of dyeability	Wool
Mixture of perflourocarbons with hydrogen	Prevention of plasticizer migration from polymer volume to surface	PVC.
CF_4, C_2F_6, SF_6	Decrease of surface friction, improvement of antistatic properties, water-repellence and resistance to weathering	Polyester

Note: PA - polyamide; PE - poly(ethylene); PP - poly(propylene); PS - poly(styrene); PVC - poly(vinyl chloride); PC - polycarbonate; PI - polyimide; PTFE - poly(tetra fluoro- ethylene); PET - poly(ethylene terephthalate); POM - poly(oxymethylene); PMMA - poly(methyl methacrylate).

Plasma treatment conditions

The basic external parameters of any plasmachemical process are the initial gas composition, its pressure and flow rate, dissipated power or discharge current and treatment time. Some data concerning the conditions of the polymer materials modification by plasma etching are given in Table 4 [7, 9, 10, 23, 32, 33, 36, 51-58]. These data indicate the range of optimum parameters is wide enough.

One must note that such a parameter of plasma system as dissipated power includes power deposited not only in the plasma zone but also in the electrode regions of discharge. The contribution of the electrode regions into the energy dissipation and into the treatment results is not investigated enough.

The wide range of plasma treatment optimum parameters is partially explained by the low selectivity of the effects being achieved. The estimation of the influence of separate parameters on the results of fabric plasma modification [59] has shown that the discharge current and treatment time are the most important parameters. The influence of gas pressure and flow rate is not great.

It has to be noted that the absence of a similarity theory for the plasmachemical systems under consideration makes it difficult to compare the results obtained in different reactors and to predict the optimum operating conditions for an industrial reactor on the base of laboratory experiments.

Table 4.

Aim of treatment	Treatment conditions
Desizing of cotton/polyester fabric	Etching in RF-discharge in air or O_2, f=13.56 MHz, p = 133 Pa, W = 200 W, t = 20 min.
Increase of adhesion properties of poly(ethylene) films	Etching in glow discharge in O_2 or in O_2/ CF_4 mixture, p = 100 Pa, W= 50 - 200W, t = 3 - 10 min.
Improvement of absorptivity and wear resistance of polyester and polyamide fibres and knitted goods	Etching in glow discharge in air, p = 0.133 - 1.33 Pa, J = 50 -100 mA, t = 50 - 60 min.
Surface activation of polymer sheets and fabrics	Etching in glow discharge in air, p = 1.33 - 266 Pa, f = 1 kHz - 40 MHz, W = 0.1 -1 W/cm^2.
Increase of adhesion properties of poly(propylene) surface	Etching in RF-discharge in O_2, p = 66 - 133 Pa, q=10 - 400 ml/min, W = 70 - 1000 W, t = 3 - 60 s.
Prevention of plasticizer migration from PVC-film volume to surface, improvement of water repellence, oil resistance and fastness to weathering	Etching in glow discharge in H_2/C_nF_{n+2} (n = 1,2,3) mixture; the gas mixture may contain additionally CO, Ar, He, CO_2, N_2, p = 6.65 - 1330 Pa.
Improvement of polyester bonding	Etching in glow discharge in Ar, N_2, He or in

	their mixtures; $p = 0.13 - 133$ Pa, $W > 100W$.
Imparting the antistatic properties, decrease of friction coefficient values	Etching or polymer film deposition in the RF-discharge in vapours of the saturated hydrocarbons, $p = 0.001 - 5$ Pa, $f = 13.56$ MHz, $W = 500W$, $t = 10$ min.
Imparting the antistatic properties to polyester parts	Etching in glow discharge in Ar, He, CO_2 followed by grafting the antistatic agents to polymer surface. $f = 100$ kHz - 13.56 Mhz, $t=1-60$ s. Conditions of treatment in noble gas plasma: $p=$ 0.133-133 Pa, $W=0.1-30$ W/cm^2 Conditions of treatment in CO_2 plasma: $p= 0.66 - 66$ Pa, $W = 0.1-5$ W/cm^2.
Increase of adhesion properties and improvement of printing the poly(propylene) surface	Etching in glow discharge in O_2, $p = 13.3 - 1330$ Pa, $t=5$ s - 15 min.
Increase in wettability of poly(methyl methacrylate) surface	Etching in RF- discharge in NH_3, $f = 13.56$ MHz, $p=13.3$ Pa, $W=1$ W/cm^3, $t=200$ s.
Increase in wettability and in drying rate of silk fabric	Etching in RF-discharge in N_2, $f = 18$ MHz, $W = 100W$, $t \leq 90$ min.
Imparting the crease-resistance and water - repellency to cellulose, the decrease of the cellulose solubility in alkaline solutions	Etching in RF-discharge in NH_3; $f=1-30$ MHz, $p= 133-665$ Pa, $W=20-80$ W, $q=1-5$ cm^3/min, $t=2$h.
Increase in wettability by dye solutions and increase of the drying rate of cotton fabric	Etching in RF-discharge in Ar, $f=13.56$ MHz, $W=40$ W, $p = 13.3$ Pa, $q=6.5$ cm^3/min, $t \leq 90$ min.
Increase in wettability of polyester fabric	Etching in glow discharge in N_2, CO_2, air, NH_3; $f=1-10$ kHz, $j=1-1.5$ mA/cm^2, $p=66$ Pa.
Increase of water receptivity and anti-soiling properties of the polyester and cotton/polyester fabric	Etching in RF-discharge in N_2, O_2, CO_2, air; $f =$ 1 MHz or $f = 13.56$ MHz, $j=0.41 - 1.52$ mA/cm^2, $p = 133$ Pa.
Desizing the polyester/cotton fabric	Etching in RF-discharge in N_2, O_2, air; $f=13.56$ MHz, $W=100-300$ W, $q=50$ cm^3/s, $t \leq$ 20 min.
Increase of the polyamide fibre adhesion	Etching in microwave discharge in Ar, N_2, NH_3 followed by grafting the allylamine to polymer surface; $f=2.45$ GHz, $W=100-700W$, $p=0.66 - 133$ Pa, $t=5-60$ s.
Decrease of the wool fabric shrinkage	Etching in RF-discharge in Air, O_2, N_2, H_2, CO_2, He, NH_3, $f = 13.56$ MHz, $W=30-100$ W, $p=266-532$ Pa, $t= 0.7 - 1.5$ s;

Note: p - gas pressure, W - dissipated power, t- time of treatment, J - discharge current, j - discharge current density, f - frequency of the external electromagnetic field, q - gas flow rate.

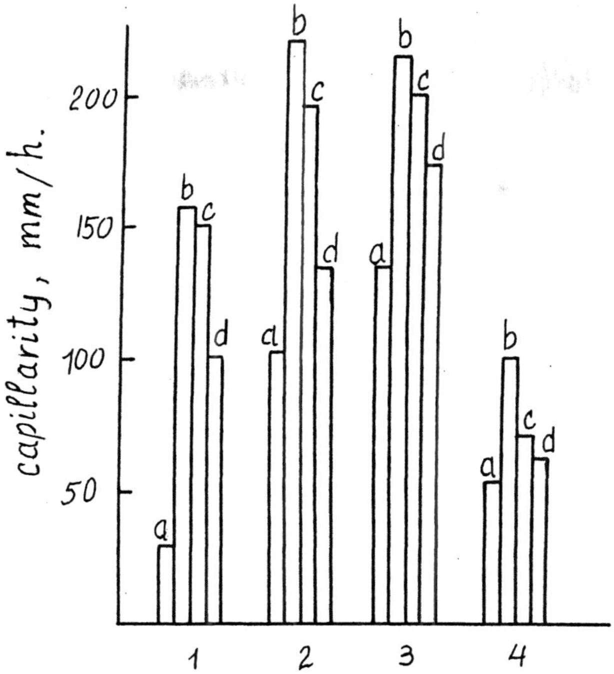

Fig. 1. The influence of heating and steaming on the capillarity of plasma- treated fabrics.
a- capillarity before treatment, b- capillarity after plasma treatment, c- capillarity after plasma treatment followed by heating, d- capillarity after plasma treatment followed by steaming;
1- "Nadezhda" cotton fabric,
2- Calico cotton fabric, 3- Shiatings cotton/polyester fabric,
4- "Polyefirnaya" polyester fabric.

Stability of plasma treatment effects

The required stability of plasma treatment results depends on the aim of the treatment. The requirements differ significantly in the case of technological and consumer aims. Experiments show the surface properties attained by plasma action to be partially reversible. Probably, it is caused by a number of factors including the adsorption phenomena, relaxation of non-stable active sites and surface charge and the processes connected with the mobility of macromolecule segments. The segment mobility leads to the removal of plasma-formed functional groups from the superficial layer into the polymer volume. Aqueous and thermal treatment influence the stability of plasma-induced effects as is shown in Figure 1 [60].

Problem of combining the plasmachemical and conventional wet processes in fabric finishing technology

In spite of numerous possibilities and advantages of plasma processes their role in the industrial manufacturing has not been envisaged as yet. Some possible applications of plasma for fabric finishing are discussed in [61-66]. One can outline two main approaches: (1) substitution of one or several conventional operations for a plasmachemical treatment, (2) combination of plasma and conventional processing. The latter is expedient if the effect of the combined action exceeds the results of the plasmachemical and conventional processes taken separately.

The advantages and sometimes the necessity of plasmachemical treatment are evident as it concerns the environmental problems, for example in the substitution of the wool fabric chlorination for plasma treatment.

On the whole, the application of plasmachemical processes in textile industries allows to substitute the liquid-phase operations for dry gas-phase techniques. Their wide application will lead to the decrease of water consumption.

At the same time the combination of liquid and dry plasma processes in fabric finishing technology can cause some problems. One of them is connected with the fact that the plasma treatment is a vacuum process, and extensive gas and water vapor evolution from material after liquid treatment is undesirable in a plasma reactor. In this case the vacuum drying of the fabric is probably needed before plasma treatment.

The problem arises how to match the plasma reactors and the conventional equipment working in the same production line. The solution of this problem depends mainly on the development of methods for the calculation of operating parameters of plasmachemical reactors to enhance their capacity. An important task is to design the high-speed plasma apparatuses of continuous action.

Industrial-scale plasmachemical reactors and processes for fabric finishing developed in Russia

Industrial application of fabric plasma treatment requires hi-tech plasmachemical equipment. The problem is to treat the great quantity of the moist material under controlled vacuum conditions and to maintain the stable uniform plasma in a reactor of large volume. This problem did not arise in microelectronics technology as the size of material under treatment is significantly less than the size of a textile cloth.

The researches performed at the NIEKMI Institute , Ivanovo, Russia resulted in the creation of a series of machines for industrial plasmachemical treatment of fabrics. The first experimental UPCH-140 machine (Fig.2) was produced in 1986 for the treatment of fabrics up to 140 cm wide. The power of glow discharge excited in the reactor of this machine is up to 60 kW, and it allows to maintain the treatment rate in the range of 4 - 40 m/min. Laboratory tests and industrial application of this machine have shown that the optimum pressure range for the treatment of fabrics is about 50 - 150 Pa and the optimum discharge current range is about 1.5 - 2.0 mA/cm^2.

The LPCH-180SH plasmachemical system has been developed for pre-treating the shawl wool fabrics before printing. The length of the material being treated in a plasma zone is 8.5 m which corresponds to the treatment rate up to 60 m/min. The KPR-180 machine (Fig. 3) is a new automatic equipment for plasma treatment of various textile materials. Some characteristics of LPCH-180SH and KPR-180 machines are listed in Table5.

Table 5.

Parameter	LPCH-180SH	KPR-180
Working width, cm	180	180
Diameter of fabric roll, cm	≤ 135	≤ 110
Treatment rate, m/min	8 - 80	8 - 80
Power consumption, kW	180	75-110
Overall dimensions, cm	1350×780×400	780×520×420
Mass, kg	25000	9000

Today some textile enterprises in Russia have an experience in the application of plasma for finishing wool and cotton fabrics. The LPCH-180SH and KPR-180 machines are in operation at the Pavlovsky Posad Shawl Production Company in Moscow region. Usually, the wool shawl fabrics pre-treatment consists in the fabric chlorination in

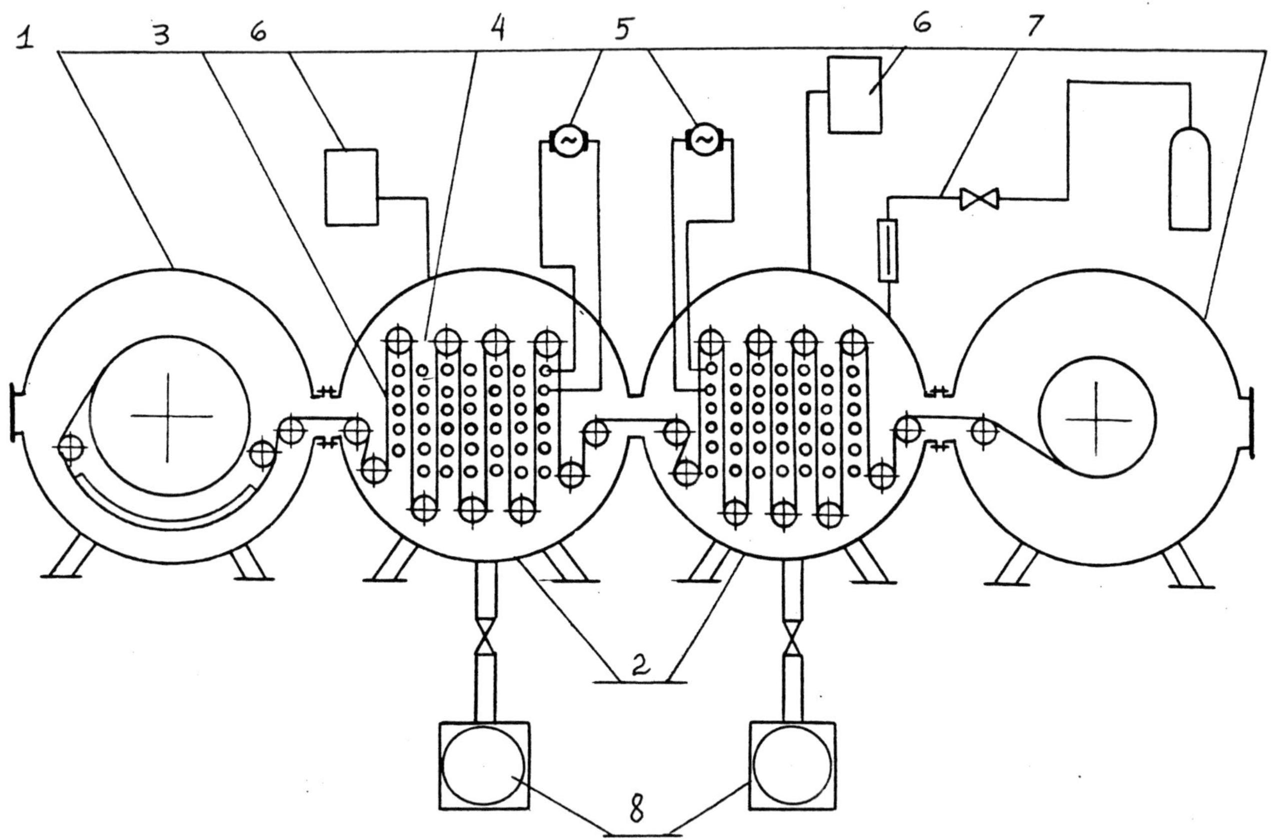

Fig. 2. Schematic diagram of UPCH-140 machine.
1- vacuum chamber, 2- plasmachemical reactor, 3- fabric, 4- electrodes, 5- power supply, 6- vacuum gauge, 7- gas supply system, 8- vacuum pump.

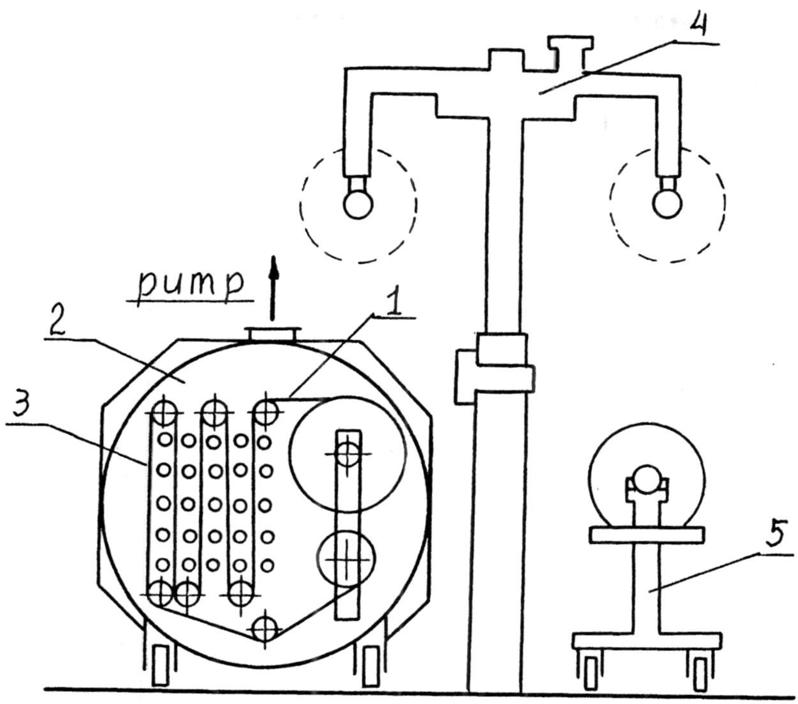

Fig 3. Schematic diagram of KPR-180 machine.
1- plasmachemical reactor, 2- fabric, 3- electrodes, 4- roll loader, 5- transporting cart.

sodium hypochlorite solution. Sodium hypochlorite is produced by the interaction of gaseous chlorine with caustic soda. The use of the gaseous chlorine causes the problems of equipment corrosion and environmental protecting. That is why the chlorination process at this enterprise is now substituted for the plasma treatment. The desized and dried fabric is treated by the air plasma. Later the fabric is printed without chlorination, stannation and washing. The plasma-treated fabrics have capillarity high enough for printing under usual conditions. Whiteness degree of these fabrics is higher than that of chlorinated wool (Table 6). This is probably a result of the more soft action of plasma on material. Moreover, the presence of ferrous salts in washing waters does not cause the yellowing of the fabric because the products of interaction of these salts with chlorine are not formed. Thus the stannation is not necessary if plasma treatment is used.

Table 6.

Properties	Chlorinated wool fabric	Plasma-treated wool fabric
Whiteness degree before printing (%)	35.8 / 38.6	45.6 / 45.1
Capillarity before printing (mm/h)	105 / 55	170 / 140
Fastness of colouring (number) to		
water	3 / 3	3 / 4
dry rubbing	4 / 4	4 / 3-4
dry cleaning	5 / 5	5 / 5
ironing with steam	5 / 5	5 / 5
Breaking load (N)		
in warp	278 / 180	298 / 200
in weft	195 / 170	210 / 195
Stiffness ($\mu N/cm^2$)		
in warp	4570 / 2120	3090 / 1980
in weft	3340 / 1800	1445 / 1110

Note: The first value is for 60p article fabric and the second value is for 71p article fabric.

The whole assortment of dyes used in shawl production was tested during the use of plasma equipment. The increase of fabric whiteness was noticed to increase the printed design brightness for such dyes as the Rhodamine G, the acid yellow light-resistant dye, the acid bright blue dye. The fastness of colouring is the same for the plasma-treated and conventionally pre-treated fabrics. Moreover, plasma treatment causes some increase in the fabric breaking load. These fabrics possess softer feel in comparison with the chlorinated wool.

Plasmachemical treatment is successfully used for finishing cotton textile materials. The UPCH-140 machine is applied at the Bolshaya Ivanovskaya Manufaktura textile enterprise (Ivanovo, Russia) for pre-treating grey or mercerized cotton, cotton/polyester and cotton/nylon blended fabrics before dyeing. According to the conventional technology, the required capillarity of these fabrics is achieved during such pre-treatments as desizing and alkaline boiling or during simultaneous fabric boiling and mercerization, or during boiling and bleaching in a rope form. These processes are laborious, time-consuming and require a lot of chemicals, steam and water. Wettability of plasma-treated fabrics is high enough to exclude the alkaline boiling process. Such a result is especially important for blended fabrics because their alkaline boiling leads to the formation of the creases. Plasma-treated fabrics can be dyed in deep shades by sulphur, active and vat dyes.

Plasmachemical technology gives the possibility to produce the new cotton/polyester fabric "Premiere" which looks like a lace. The semitransparence of this fabric is achieved as a result of the destructive action of special chemicals on a cotton component of the yarn. The chemicals are applied to the mercerized and plasma treated fabric by means of a printing machine followed by thermal treatment. The products of cotton destruction together with the excessive chemicals are removed during the alkaline boiling and bleaching of the fabric. A high quality lace-like material cannot be produced without the plasma treatment because the required complete removal of the destructed cotton is not achieved.

Industrial applications of the plasma treatment for fabric finishing require the solution not only of the specific technical problems concerning the design of the plasmachemical equipment, but also the solution of some fundamental problems.

The plasma in an industrial reactor is completely restricted by the material under treatment which represents a chemically reacting boundary surface. Therefore the quantitative description of such a plasma should take into consideration the boundary conditions not only for an electromagnetic field and the charged particles but also the conditions for the chemically active species. Thus we have to know the kinetic characteristics of the interaction between these species and the polymer materials under treatment. It is also necessary to take into consideration the transport of the active particles in a volume of the fabric which is considered to be a capillary-porous body. The data on kinetic characteristics of active species interaction with polymers are rather limited in literature and the transport phenomena of the plasma active particles in fabrics have not been investigated.

The most complicated and important problem is the influence of the gaseous products on the plasma properties. The fact is the quantity of gaseous products of plasma interaction with polymers in an industrial reactor is so great that we are principally unable to speak about the plasma in the pure initial gas. The data about the properties of plasmas in pure gases obtained in the laboratory experiments can not be used for the analysis of the plasma properties in an industrial reactor. The chemically reacting plasma must be considered to be a self-consistent, self-organizing system which includes not only the gas phase but also chemically reacting polymer surface layer. Theoretical and experimental investigations of such plasma are not available. Significant progress in the development of new complicated plasmachemical processes such as a synthesis of the surface layers with the required properties is impossible without understanding the interconnection between plasma-induced chemical transformations and the physical plasma properties of system.

References

1. W. Rakowsky, M. Okoniewsky, K. Bartos and J. Zawadsky, *Melliand Textilberichte,* (4) (1982) 307.

2. Patent Certificate 4536274, USA.

3. Y. Yriyama, H. Yasuda, *Polym. Mater. Sci. and Eng.* **56**: Spring Meet., Denver, Colo, 1987.-Washington, D.C. (1987) 327.

4. Patent Application 59-86635, Japan.

5. Patent Certificate 4722816, USA.

6. Patent Certificate 460945, USA.

7. Patent Application 61-37826, Japan.

8. Patent Application 61-20123, Japan.

9. Patent Application 59-51921, Japan.

10. Patent Application 61-64734, Japan.

11. S.F. Sadova, L. Lapchik and O. Paigrt, *Izv. Vuzov. Technol. Tekst. Prom.* (5) (1983) 53-56 (in Russian).

12. S.F. Sadova, L. Lapchik , O. Paigrt and V. Kalauskova, Ibid, (6) (1983) 64-68 (in Russian).

13. M.V. Akulova, I.B. Blinicheva, B.N. Mel'nikov and A.I. Maximov *Izv.Vuzov. Khimija i Khim. Tekhnol.* **25** (11) (1982) 1376-1379 (in Russian).

14. T. Wakida et al., *J. Soc. of Fiber Sci. and Technol., Japan,* **42** (2) (1986) 69-73.

15. I.B. Blinicheva, M. Yu. Burmistrova and B.L. Gorberg, *Izv. Vuzov. Tekhnol. Text. Prom.* (12) (1988) 141-146 (in Russian).

16. A. Wrobel et al., *Polymer,* **19** (8) (1978) 908-912.

17. M. Rasche, *Adhasion,* **30** (3) (1986) 25-28.

18. Yu. V. Lipin, A.V. Rogachev and V.V. Kharitonov , The vacuum metal deposition on polymer materials (Leningrad: Khimiya, 1987) (in Russian).

19. Yu. V. Lipin et al., The preparation of the polymer film surface to the vacuum metal deposition (Riga: LatNIINTI, 1990) (in Russian).

20. R.L. Bersin, *Polymer News,* **11** (11/12) (1976) 13-18.

21. L.M. Anischenko, S.B. Kuznetsov and S.Yu. Lavrenyuk. *Fizika i Khimiya Obrabotki Materialov*, (6) (1985) 124(in Russian).

22. C. Haag and H. Suhr, *Appl. Phys. A: Solids and surfaces*, **7** (2) (1988) 199-203.

23. L. Dorn, I. Gartner and M. Rache, *Kunststoffe*, 76 (3) (1986) 249-253.

24. V.A. Zharov and O.N. Solovyova, *Electronnaya obrabotka materialov,* (5) (1986) 49-51.

25. Patent Application 59-56430, Japan.

26. S.J. Kaplan and P.W. Rose, *Plast. Eng.* **44** (5) (1988) 77-79.

27. L. Mascia, G.E. Corr and P. Kember, *Plast. and Rubber Process. and Appl.* **9** (3) (1988) 133-140.

28. Patent Certificate 1062323, USSR.

29. J.E. Koresh, S.S. Barton and M.J. Evans, *Carbon*, **24** (3) (1986) 325-330.

30. Patent Application 2155025, UK.

31. Patent Application 62-91535, Japan.

32. Patent Application 61- 157536, Japan.

33. Patent Certificate 4351857, USA.

34. K. Nakamae, K. Yamaguchi, S. Tanigava and K. Sumiya , *J. Chem. Soc. Jap., Chem. and Ind. Chem.* (11) (1987) 1995-2000.

35. R. E. Allred, E. W. Merrill, and D. K. Roylance, *Amer. Chem. Soc. Pol. Prepr.* **24** (1) (1983) 223-224.

36. M.R. Wertheimer, and H.P. Schreiber, *J. Appl. Polym. Sci.* **26** (6) (1981) 2087-2096.

37. Patent Certificate 4560458, USA.

38. Patent Application 59-30832, Japan.

39. Patent Application 63-35632, Japan.

40. M. Strobel, S. Corn, C.S. Lyons and G.A. Korba, *J. Polym. Sci.: Polym. Chem.* **A25** (5) (1987) 1295-1307.

41. M. Kagoma et al., *J. Phys. D.: Appl. Phys.* **20** (1) (1987) 147-149.

42. J. Momose, T. Takada and S. Okazaki, *J. Appl. Polym. Sci.: Appl. Polym. Symp.* (42) (1988) 147-149.

43. I.H. Loh, M. Klausher, R.F. Baddour and R.E. Cohen, *Polym. Eng. and Sci.* **27** (11) (1987) 861-866.

44. R. Chasset, G. Legeay, J.C. Touraine and B. Arzur, *Eur. Polym. J.* **24** (11) (1988) 1049-1055.

45. M. Strobel et al., *J. Appl. Polym. Sci.: Appl.Polym. Symp.* (42) (1988) 73-96.

46. Patent Certificate 450878, USA.

47. Patent Application 3408837, Germany.

48. M. Strobel, S. Corn and C.S. Lions, *J. Polym. Sci.: Polym. Chem. Ed.* **23** (4) (1985) 1125-1137.

49. Patent Application 61-60730, Japan.

50. Patent Application 61-133239, Japan.

51. G. Bogoeva-Gaseva, D. Losinski and G. Petrov, *Polymer*, **7** (4-5) (1986) 103-104.

52. Patent Application 59-126437, Japan.

53. Y. Osada and Y. Iriyama, *Thin Solid Films*, **118** (2) (1984) 197-202.

54. Patent Application 61-237, Japan.

55. Patent Application 50-248742, Japan.

56. Patent Application 59-98141, Japan.

57. Patent Application 61-171740, Japan.

58. B.L. Gorberg, A.I. Maximov and B.N. Mel'nikov, *Izvestiya Vuzov. Khimiya i Khim. Tekhnol.* **26** (11) (1983) 1362-1376.

59. B.L. Gorberg, *Modification of Textile Materials in Low-temperature Plasma of Glow Discharge* (Dissertation for the degree of Candidate of technical sciences, Ivanovo, 1985) (in Russian).

60. L.V. Sharnina, Development of the effective methods for the plasma activation of textile materials(Dissertation for the degree of Candidate of technical sciences, Ivanovo, 1990) (in Russian).

61. B.N. Mel'nikov, I.B. Blinicheva and A. I. Maximov, *Prospects of the Plasma Technology in Textile Industry* (Moscow, 1985) (in Russian).

62. *Progress in the Textile Chemistry*, Ed. B.N. Mel'nikov (Moscow: Legprombytizdat, 1988) (in Russian).

63. T.S. Novoradovskaya and S.F. Sadova, *Chemistry and Chemical Technology of Wool* (Moscow: Legpronbytizdat, 1986) (in Russian).

64. I.N. Belyaev and E.A. Rasskazova, *Modification of the Wool and Chemical Fibres in Low-temperature Plasma* (Moscow: ZNII Shersty, 1983) (in Russian).

65. *Vacuum deposition of the coatings: Collected articles* (Riga: Zinatne, 1986) (in Russian).

66. A.I. Maximov, *Vacuum Plasmachemical Technologies, in Collected Articles: Plasmachemical Technology* (Novosibirsk: Nauka, Sib. otd., 1991) 336-372 (in Russian).

The Institute of Chemistry of Non-Aqueous Solutions of the Russian Academy of Sciences, Ivanovo, Ivanovo State Academy of Chemistry and Technology

RADIATION-CHEMISTRY PROCESSES FOR THE FINISHING AND DYEING OF TEXTILE MATERIALS

B.N.Mel'nikov, M.K.Shirokova, V.G.Vermul,
V.D.Orechov, K.D.Pismannic

The results of research into radiation-chemistry processes taking place during the finishing and dyeing of textile materials with mixtures of vinyl monomers and methylolamides are presented. The highly-effective radiation-chemistry methods created by the authors for textile material finishing and dyeing are described.

One of the promising directions for the development of applied radiation chemistry is the use of irradiation in the chemical processing of fibrous materials, particularly, at the stage of material finishing [1]. The application of high energy particles for treating textile materials is attracting more and more interest due to its low energy costs, the high fixation rates of chemical substances at low temperatures, and the absence of problems connected with sewage purification. The last of these factors ensures the creation of textile material dyeing and finishing processes with fewer operations.

The heterogeneous systems "cellulosic polymer - aqueous solution of a mixture of vinyl monomers and methylolamides" were used as the focal point of our investigations. In several experiments, an aqueous solution also contained a dye. Among the variety of known methylolamides, the following preparations were chosen: the methylol derivatives of melamine (metazin, metazin 6Y), dimethylol- ethylenurea (carbamol TSEM), and dimethyloldioxyethylenurea (carbamol GL). The concentration of the methylolamides was varied from 0 to 200 g/l. The amides of acrylic and methacrylic acids, and acrylic acid itself were used as co-reagents. The amount of vinyl monomers in the solutions was altered from 0 to 100 g/l. The dyeing of a textile material was performed using azo-, antraquinone, and phthalocyanine dyes (up to 20 g/l). The irradiation of a system "textile material - aqueous solution of finishing agents and a dye" was done at 25-30°C. A ELT-

1.5 electron accelerator (producing electrons with energy up to 1.5 MeV) and a YE 04Ì-600 accelerator (producing electrons with energy up to 0.4 MeV) were used as the sources of quick electron flow. The ionizing radiation dose was equal to 10-15 kGy, the dose rate being 50 krad/s. The textile material subjected to irradiation was treated with heat at 100-150°C for 3-7 minutes.

The character of the changes in the mixture of vinyl monomers and methylolamides (located in cotton fiber) was defined according to the total nitrogen content and the content of different formaldehyde forms (which are educed after the selective resin hydrolysis and the hydrolysis of all the cross-links, methylene bridges ($>NCH_2N<$), and $>NCH_2OR$ - bonds). The quantitative analysis of formaldehyde forms was made in conformity with procedures described in [2-4]. The total nitrogen content was evaluated using the Kj'eldal semimicromethod [5]. The amount of nitrogen in melamine was defined by UV-spectroscopy [6]. The portion of nitrogen belonging to the acrylamide links was calculated as a remainder obtained after the subtraction of melamine's nitrogen from the total nitrogen amount. The number of cross-links between the cellulose macromolecules was estimated using Frick's formula [7]. In these calculations, only the nitrogen corresponding to the links of melamine, ethylenurea, etc. was taken into account. The degree of linkage of cotton cellulose was judged by cellulose swelling and its solubility in a cuprammonium solution. This solubility was defined using Mehta's method [8]. The degree of fixation of finishing agents was calculated as a ratio of fast-fixed resin (which remains in the fabric after the removal of water-soluble products created during the reaction of resin formation) to the total amount of resin on the fabric [9].

Hydrolytic removal of the products of melamine transformation from the fabric was performed using an aqueous solution containing 5% of urea and 1.5% of orthophosphoric acid at 80°C for 60 minutes [10]. The IR-spectra of modified cellulose were obtained using two spectrophotometers: IR-20 and Specord-75 IR.

The composition of the resin formed in the fiber during the application of the radiation-chemistry method of the initiation of reactions has been studied. On the basis

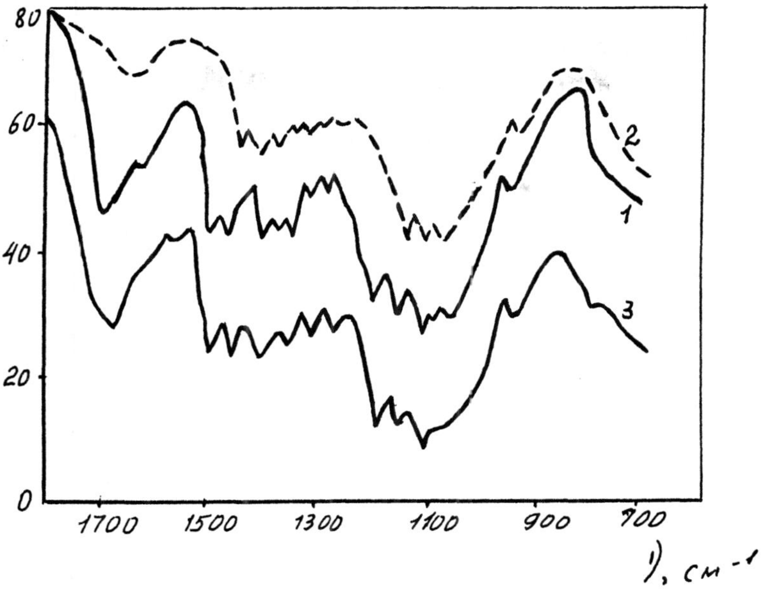

Fig.1. IR-spectra of cellulose modified with acrylamide using radiation-chemistry (1, 3) and thermocatalytic (2) methods:

1 - before re-precipitation; 2, 3 - after precipitation from the cuprammonium solution.

X = ν, cm^{-1}; Y = Transparency, %

e of the data obtained from the IR-spectroscopy and from the chemical analysis of modified cellulose, it has been shown that in the absence of an acidic catalyst, high-nergy particles cause the vinyl monomer transformation accompanied by the inoculation of radical-active monomer and its homopolymerization (Table 1, Fig.1). Among the mentioned chemical processes, the formation of inoculated co-polymers is the predominating one.

Table 1.

The content of nitrogen in fabric modified with acrylamide

Acrylamide concentration in solution, g/l	The conditions of heat treatment		Nitrogen content in scoured fabric, %	
	Temperature, ^0C	Time, min	Subjected to irradiation (10 kGy)	Not subjected to irradiation
60	100	7	0.64	0.034
60	120	3	0.73	0.022
60	150	3	0.73	0.022

As follows from the data in Table 1, the fabric treated with a flow of quick electrons contains 20-80 times more fast-fixed polyacrylamide than fabric subjected to thermocatalytic modification. The experimental fact that polyacrylamide can not be removed from the cotton fiber reprecipitation produced from the cuprammonium solution (Fig.1) can serve as a proof of the formation of the covalent bond between cellulose macromolecules and polyacrylamide. In the spectrum of re-precipitated cellulose modified in accordance with the radiation-chemistry method, the bond remains (1680 cm^{-1}) characteristic of polyacrylamide functional groups (Fig.1, curve 3). When the chemical method for the initiation of acrylamide transformation reactions is applied, only the band corresponding to the molecules of adsorbed water is present in the spectra of re-precipitated cellulose in the interval of 1700-1500 cm^{-1} (Fig.1, curve 2). Being based on the obtained experimental data and also on the results of publications [11, 12], it is possible to present schematically the process of the formation of inoculated co-polymers of cellulose and acrylamide under the influence of ionizing radiation during liquid-phase cellulose modification with vinyl monomers:

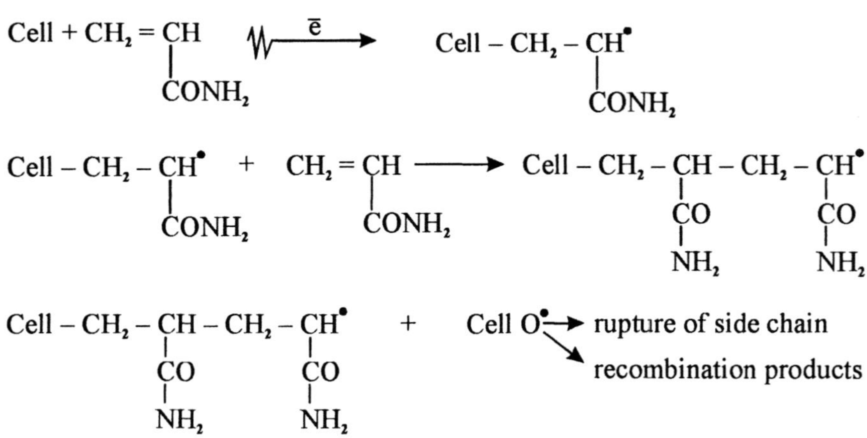

The composition of the co-polymer of cellulose with polyacrylamide has been calculated. At the maximal monomer concentration (100 g/l), the portion of cellulose, having participated in the reaction of inoculated co-polymerization, lies within the range of 5-8%.

The molecules of the methylol derivatives of urea, dioxyethylenurea, ethylenurea, and melamine, being subjected to the influence of the flow of quick electrons, to some extent react with each other and with OH-groups of cellulose.

If the radiation dose equals 10 kGy and if the treatment is performed at low temperatures (100 ^{0}C), the residues of the amide of acrylic acid predominate in the resin thus obtained (Table 2).

The UV-spectrophotometric definition of melamine in resin composition has shown that the melamine content is approximately 8 times lower than the number of acrylamide

Table 2.
The content of the resin formed in the fabric treated according to the radiation-chemistry method in the absence of a catalyst

The conditions of heat treatment		Nitrogen content corresponding to (%)		CH_2O/ N- ratio	Nitrogen content corresponding to (%)		CH_2O/ N- ratio
		melamine	acrylamide		melamine	acryl amide	
Tempera-ture, ^{0}C	Time, min	The fabric subjected to radiation			The fabric not subjected to radiation		
100	7	0.16	1.36	0.05	0.1	0.0	0.5
150	3	1.30	1.35	0.25	0.3	0.2	0.6

links. Due to the same cause, the ratio of the amount of formaldehyde to the amount of nitrogen at low temperatures of the reaction does not exceed 0.05 (Table 2). The joint influence of the acidic catalysis and of high-energy particles results in the acceleration of interaction reactions between the thermoreactive precondensate molecules and between these molecules and OH-groups of cellulose. In the presence of the acidic catalyst, together with the general intensification of the resin formation process, the content of methylolamides in resin composition is sharply increased. The molar ratio of formaldehyde to nitrogen content reaches 0.2-0.4 (Table 3).

The main part of the finishing agent participates in the reaction of cellulose macromolecule linking accompanied by the formation of cross-links. The number of cross-links for 100 anhydroglucose residues attains 3-4. The amount of cellulose

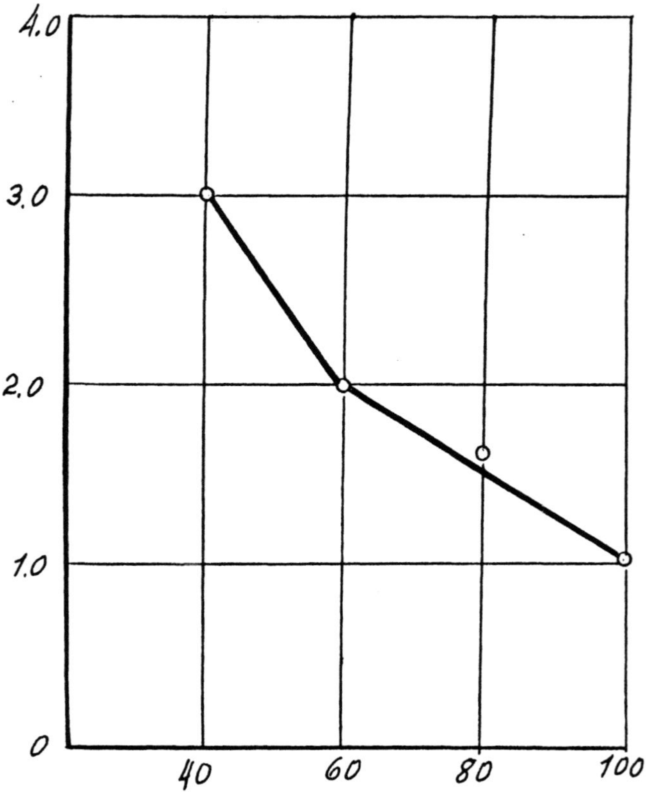

Fig.2. The dependence of the degree of cotton cellulose cross-linking upon the concentration of acrylamide.
X = Acrylamide concentration, g/l;
Y = The number of cross-links for 100 anhydroglucose residues

dissolving in the cuprammonium solution after cotton modification with the mixture of finishing agents in the conditions of radiation-chemistry treatment lies between 16.7-24.0%. The substitution of aluminum chloride for ammonium chloride results in an increase in the solubility of modified fiber up to 44 - 46%. In these conditions, the presence of vinyl monomers in the system "cellulose - methylolamide" causes the alteration of the structure and density of the cross-links favoring the formation of elastic "sewings". The redundant content of radical-active additions (for example, acrylic acid amide - 60 g/l) abruptly diminishes the degree of cross-linking of the adjacent cellulose chains from 3 to 1 cross-links for 100 anhydroglucose residues (Fig.2).

Table 3.
The dependence of nitrogen and formaldehyde content in the fiber modified by the radiation-chemistry method upon the nature of the acidic catalyst and of the finishing agent

Catalyst		The conditions of heat treatment		The total content of		CH_2O/N-ratio	The number of cross-links for 100 anhydroglucose residues after radiation-chemistry treatment
Chemical formula	Concentration, g/l	Temperature, ^0C	Time, min	nitrogen	formaldehyde		
Melamine-formaldehyde precondensate							
NH_4Cl	4	100	7	1.7	0.7	0.2	-
		150	3	2.0	1.2	0.3	2.0
$AlCl_3$	20	100	7	2.2	1.6	0.3	-
		150	3	3.5	1.9	0.3	4.0
Dimethylolethylenurea							
$AlCl_3$	20	150	3	1.4	1.1	0.4	1.5
Dimethyloldioxyethylenurea							
$AlCl_3$	20	150	3	1.17	1.05	0.4	1.5

The quantity of cellulose which does not dissolve in the cuprammonium solution decreases from 88 down to 68%. This is caused by the partial deactivation of active groups of methylolamides due to the formation of the polymeric form of acrylic acid amide resulting in the reduction of the possibility of methylolamide interaction with cellulose hydroxyl groups. The supposed interactions can be presented in the following way (R - melamine residue):

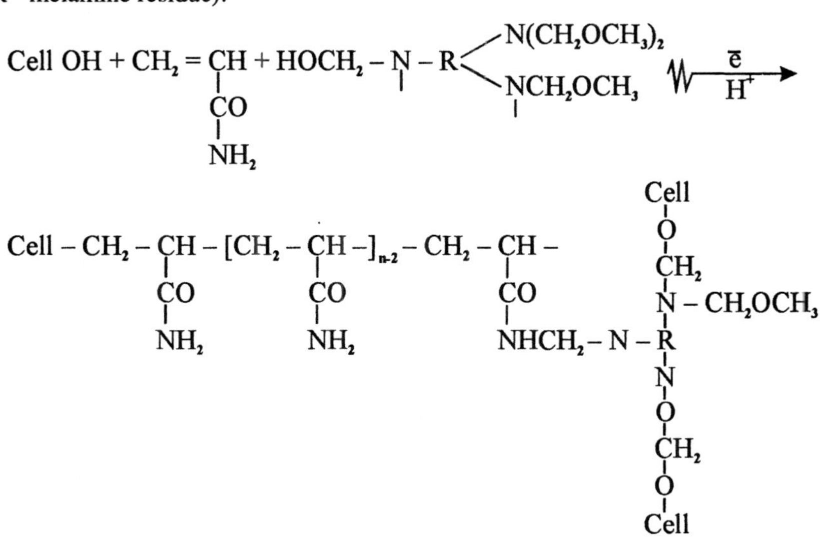

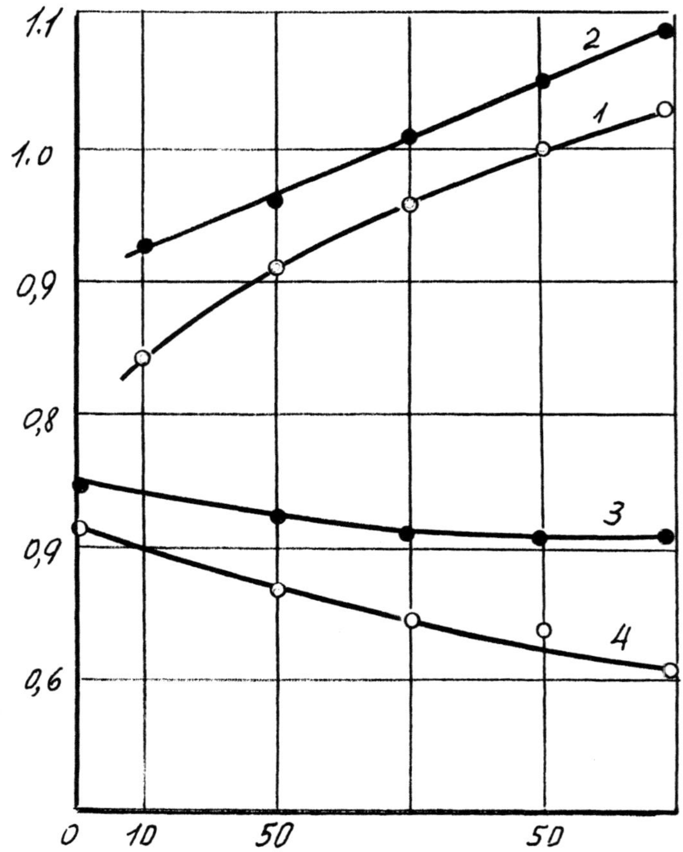

Fig.3. The influence of the radiation dose on the content of nitrogen connected with acrylamide (1, 2) and with melamine (3, 4):
1, 3 - 40 g/l of nitrogen; 2, 4 - 60 g/l of nitrogen.
X = Radiation dose, kGy; Y = Nitrogen content, %.

The composition of the resin thus formed and the structure of cross-links between the adjacent cellulose chains in the irradiated system "cellulose - vinyl monomer - methylolamide" partially depends upon the dose of ionizing radiation. The separate definition of the content of nitrogen belonging to acrylamide and melamine residues indicates that when the dose of radiation is increased from 10 to 50 kGy, the portion of nitrogen identified with acrylamide residues rises from 0.9 to 1.1% (Fig.3, curves 1,2), whereas the content of melamine nitrogen reduces from 0.75 to 0.59% (Fig.3, curves

3,4). Further, the ratio of formaldehyde to nitrogen content diminishes from 0.30 down to 0.24-0.25. It has also been established that the radiation-chemistry method for the initiation of chemical reactions (in comparison with the thermocatalytic method) causes a more complete transformation of finishing agents and a higher degree of polymerization and condensation (Table 4).

Table 4.
The dependence of the degree of fixation of finishing agents upon the applied method of the initiation of chemical reactions

Applied method for the initiation of chemical reactions	The degree of fixation of finishing agents, %	The degree of acrylamide fixation, %	Formaldehyde content (%) corresponding to $>NCH_2OH$ - groups
Radiation-chemistry	90-100	100	0.07-0.26
Thermocatalytic	84-87	80	0.20-0.30

The fixation degree of finishing agents in the fiber equals 90-100% when the radiation-chemistry method is used and only 84-97% after thermocatalytic fabric treatment. The degree of acrylamide transformation reaches 100% for the system which has not been subjected to irradiation.

Among the products of resin formation, the amount of formaldehyde corresponding to the free (unlinked) methylol groups is less for 10-20% when radiation-chemistry method is used in comparison with the thermocatalytic one. The intensification of the processes of resin formation and the cross-linking of cellulose macromolecules with N, N'-methylolamides in the acidic medium under the influence of the flow of quick electrons is accompanied by an increase in dye fixation degree. The highest yield of fast-fixed dye (93.6-99.4%) is achieved during fabric dyeing with reactive, direct, vat, and acid dyes (Table 5).

Table 5.
The fixation degree of dyes and finishing agents in fabric treated by the radiation-chemistry method

Dye	Fixation degree of formaldehyde-containing products %	The degree of in completion of the reaction at methylol groups, %	Dye fixation degree %
Reactive (C.I.):			
Violet 5 (18097)	93.5	10.5	98.9
Yellow 2 (18972)	96.5	7.5	95.8
Blue 76	92.5	6.5	98.0
Direct (C.I.):			
Yellow (no equivalents in C.I.)	92.5	8.5	97.2
Blue 1 (24410)	96.5	10.0	99.4
Acid (C.I.):			
Yellow 11 (18820)	93.0	9.2	94.0
Blue 83 (42660)	9275	9.4	93.6
Vat (C.I.):			
Yellow 4 (59100)	96.5	10.5	98.5
Antraquinone blue (no equivalents in C.I.)	93.0	9.2	97.2

The type of acrylic addition noticeably influences the degree of dyestuff fixation in the textile material. The degree of fixation of C.I Reactive Violet 5 (18097) and of C.I. Direct Blue 1 (24410) in cotton fiber (modified with a mixture of acrylic and methacrylic acids with metazin 6Ó) is equal to 95.1-100% and to 96.4-9.6%, respectively. The introduction of acrylic acid into the dyeing solution containing the finishing agent causes a reduction in the amount of fast-fixed reactive dye, C.I. Violet 5, down to 81.7-82.6%. In comparison, the fixation degree of the direct dye, C.I. Blue 1, equals 70.2-84.0%. (Table 6).

Table 6.
The dependence of dye fixation degree upon the nature of the acrylic derivatives used

Acrylic derivative	Cross-linking agent	Dye	Dye fixation degree, %
Acrylic acid Methacrylic acid Acrylamide Methacrylamide	Metazin 6 Ó	C.I. Reactive Violet 5	82.6 81.2 95.1 95.3
Acrylic acid Acrylamide Methacrylamide	Metazin 6 Ó	C.I. Direct Blue 1	70.2 99.2 99.2
Acrylic acid Acrylamide	Carbamol ÖÝÌ	C.I. Reactive Violet 5	81.7 100.0
Acrylic acid Acrylamide	Carbamol ÖÝÌ	C.I. Direct Blue 1	84.0 96.4

Table 7.
The indices of the coloring's fastness of fabrics, treated according to one-stage radiation-chemistry (1) and thermocatalytic (2) methods, to hydrolysis

Dye	The content of dye removed from the fabric by hydrolysis			
	pH = 3		pH = 5	
	1	2	1	2
C.I. Reactive Red 2	66.6	85.5	48.6	78.2
C.I. Reactive Orange 13	55.7	72.1	47.9	73.9
C.I. Reactive Violet 5	31.6	72.1	26.3	59.2

The results of the quantitative estimation of the solubility of modified cellulose in the cuprammonium solution confirm that under the influence of high-energy particles, the dyes take an active part in the interactions not only with thermoreactive resin (formula 1), but also with native cellulose macromolecules and with the co-polymer of cellulose and acrylamide (formula 2):

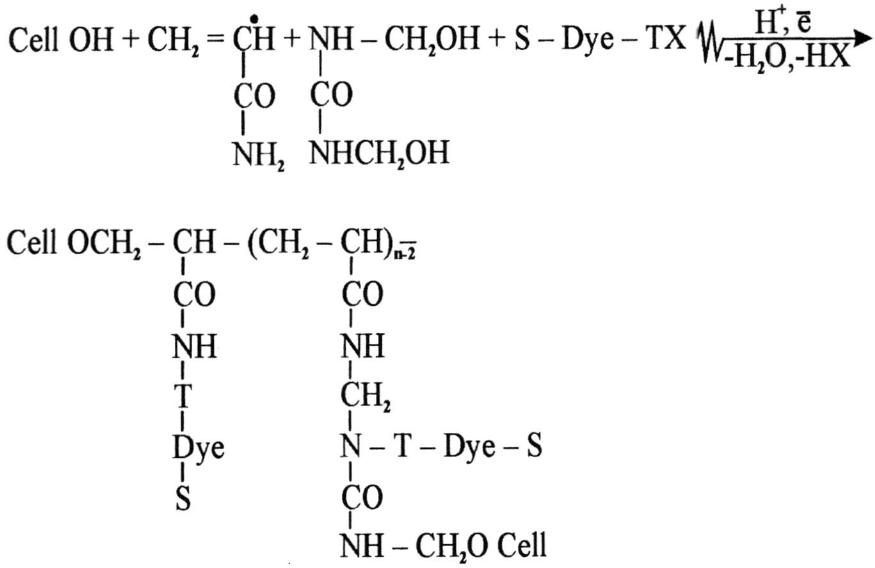

(1)

$$\text{Cell H} + \text{XT} - \text{Dye} - \text{S} \xrightarrow{\bar{e}} \text{Cell T} - \text{Dye} - \text{S} \qquad (2)$$

The revealed reactions have a substantial influence on the dye distribution between the resin and the fibrous polymer. This influence consists in increasing the portion of dyestuff which is fixed directly in the cellulosic polymer. Kinetic research into the hydrolysis of the bonds "cellulose - resin" and "cellulose - resin - dye" has shown that radiation-chemistry treatment promotes the improvement of fastness properties of the resin being formed and also of the colorings towards hydrolytic action of acidic and alkaline media (Table 7).

Radiolysis processes have been investigated which take place in the system "cellulose - finishing agent - dye", as well as the changes in the reaction ability of cellulose and finishing agents subjected to irradiation. It has been noted that the presence of radical-active monomers of the finishing agents in the fabric slows down the process of cellulosic polymer radiolysis during irradiation. These can be obviously retraced while varying the radiation dose within an interval of 5-100 kGy. In the analyzed irradiation conditions, in the absence of free radical's acceptors, the losses of tensile strength of bleached coarse calico (article 276) reach 15-37%. In the presence of melamine-formaldehyde precondensate, the decrease in tensile strength of the irradiated fabric does not exceed 2.5-21.5%. The reduction of tensile strength of the irradiated fabric containing methacrylamide lies within the range 0-12.5%. The process of cellulosic fiber radiolysis which causes the rapid diminution of the molecular mass of cellulosic polymer is accompanied by the generation of additional amounts of carbonyl and carboxyl groups in the irradiated cellulose. After the radiation treatment of fibrous material (containing a finishing agent) using a dose of 100 kGy, the resin content does not change and remains equal to 100%.

The resistance of reactive, acid, direct, and vat dyes to irradiation has been evaluated. It has been established that in the conditions studied, the most resistant are the chromophoric systems of antraquinone and vat dyes, whereas the chromophoric systems of azo-dyes are characterized by having the least resistance. It must be noted that at a constant power of irradiation, the radiolysis of the investigated dyeing systems proceeds more intensively when the radiation doses are higher than 20-30 kGy. The kinetic parameters of dye destruction under the influence of irradiation have been calculated. The increase of the drying temperature of irradiated fabrics up to 120 and 150 ^{0}C leads to the higher values of destruction rate constants, especially when using those dyes that are less resistant to irradiation (Table 8).

The influence of ionizing radiation on the reaction ability of reactive, direct, acid, and vat dyes has been established. The alteration of the radiation dose from 0 to 100 kGy gives changes in the fixation degree of bromacrylamide dye (Lanasole Blue «Ç» - no equivalents in C.I. available) and vinylsulfonic dye (C.I. Reactive Violet 5) equaling 0-34.2% and 0-3.7%, respectively. If the drying temperature of the fabric is raised, the dye fixation degree increases. The degree of fixation of direct and acid dyes (achieved after the radiation treatment) lies within the range of 3.5-15% while the degree of fixation of vat dyes reaches 8-15%.

It has been found that the initiation of the chemical reactions in the system "cellulose - finishing agent - dye" by the application of quick electron flow can influence the color characteristics of dyed fabric. The least change in the color characteristics of dyed fabrics is observed in the presence of melamine-formaldehyde precondensate or its mixtures with methylol derivatives of urea, ethylenurea, or dioxyethylenurea. It has been proven experimentally that the radiation-chemistry method of dyeing and finishing enables the production of dyed fabric with a similar, or even higher, color fastness to light than the thermocatalytic method.

Table 8.
Activation energies (E) and irradiation destruction rate constants (K_1) of reactive, acid, and direct dyes

Dye (C.I.)	Irradiation interval, Mrad	Conditions of thermotreating	K_1	E, J/mol
Reactive	1-5	-	0.18	
Violet 5	0-3	120^0C, 3 min	0.29	461.20
	0.5-3	150^0C, 3 min	0.32	
Reactive Red 2	0-3	120^0C, 4 min	0.43	-
Lanasole Blue	0-3	-	0.20	
«Ç»	0.5-3	120^0C, 4 min	0.30	3099.60
(no equiv.in C.I.)	0.5-3	150^0C, 4min	0,58	
	0-3	-	o.17	
Acid Blue 83	0-3	120^0C, 3 min	0.25	1305.70
	0-3	150^0C, 3 min	0.33	
Acid Red	0-3	-	0.23	
(analog in C.I. is	0-3	120^0C, 3 min	0.34	599.80
absent)	0-3	150^0C, 3 min	0.36	

The discussed peculiarities and advantages of the radiation-chemistry method for the initiation of chemical reactions give the possibility to create radiation-chemistry processes for cotton fabric finishing and dyeing.

The recommended operations for the technological implementation of the method developed by the authors includes the following:

– textile material impregnation in an aqueous solution containing amino-aldehyde precondensate, vinyl monomer, and a catalyst;

– squeezing, irradiation by a flow of fast moving electrons and drying with hot air at temperatures not higher than 120^0C for 3-4 minutes, followed by a slight washing.

The last of the encountered operations is necessary when melamine-formaldehyde precondensate is used as the finishing agent. If dimethyloldioxy- ethylenurea is used as the cross-linking agent and ammonium chloride (or the mixture of ammonium chloride and magnum chloride in a mass ratio of 1:1) as the acidic reagent, washing of the treated fabric is not required.

The generalized data of the physicochemical testing of treated fabric are presented in Table 9.

The test trials for the created radiation-chemistry method of crease-resistant low-shrink finishing of cotton fabrics were performed at the experimental production unit at the Glukhov cotton group of enterprises and on an experimental-industrial line in the former GDR.

In comparison with the thermocatalytic method which is the most widely used at present, the radiation-chemistry process for the crease-resistant treatment of cotton fabrics (which has been developed by the authors) possesses the following advantages:

–a substantial lowering in the quantity of chemical reagents used;

–a much shorter processing time for the treatment of textile materials within the whole finishing cycle;

–a two-fold reduction in the quantity of energy required;

–a large decrease in water expenditure;

–a reduction in equipment size.

The combined radiation-chemistry method of dyeing and crease-resistant finishing of cellulosic textile materials has also been created. It presupposes the impregnation of fibrous material with a dyeing-finishing composition containing a finishing agent, an acidic catalyst, and an organic dye.

It is expedient to use the following preparations as finishing agents: methyl ester of penta- or hexa-methylolmelamine (metazin, metazin 6Y), dimethylol- ethylenurea (carbamol TSEM), dimethylolurea (carbamol), and mixtures of the cited agents with the amides of unsaturated acids. Among the organic dyes, reactive, direct, acid, vat, and disperse dyes are suitable for the developed combined method.

In order to check the main parameters of the combined process of crease-resist finishing and dyeing of cotton fabrics, several types of cotton fabric were treated according to this process in the industrial conditions of the Glukhov cotton group of enterprises. The results of the industrial trials are presented in Tables 10 and 11.

It has been shown that the advised radiation-chemical method of dyeing and crease-resist finishing of textile materials allows the production of shirting with high color indices and with consumer properties fully corresponding to the adopted requirements for cotton fabrics with crease-resist treatment (GOST 17504-80). The application of the discussed method leads to the reduction of tensile strength losses down to 11.7-24.0% in the warp and down to 22.8-35.95 in the weft. At the same time, the effect of crease-resistance is rather high, and the total angle of fabric crease recovery lies within 220-276 degrees. In addition, such treated fabrics acquire shrink resistance. The changes in linear dimensions do not exceed 0.6-1.0% in warp and 0.1-1.8% in weft. Due to the high fixation degree of dyes and finishing agents (89.1-100%), a marked decrease is observed in the concentrations of dyestuffs and formaldehyde-containing products in the waste water.

Table 11.
The indices of the fastness of the colorings obtained on cotton fabrics treated in industrial conditions

Fabric title, article	Dye (C.I.)	Acrylic addition	Color fastness (number)			
			soaping at 40^0C	perspiration at 45^0C	dry rubbing	wet rubbing
"Liana", 966	Reactive Yellow 1	Emukryl M	5/5/5	5/5/5	5/5	5/5
"Talka", 986	Reactive Green 8	Methacryl-amide	5/5/5	5/5/5	5/5	4/4
	Direct Blue 1	Emukryl M	4/5/4	4/5/4	4/4	4/4
"Nadezhda", 1511	Reactive Green 8	Methacryl-amide	5/5/5	5/4/5	5/5	5/4
	Direct Blue1	Emukryl M	4/5/4	3/5/3	4/5	4/4

In comparison with the known combined methods of cotton fabric dyeing and crease-resistant finishing, the developed radiation-chemistry process gives the opportunity to decrease the quantity of vapor (by 26.3%), water (by 13.6%), and electric energy (by

17.4%); to reduce the plant area (by 14.6%) and specific capital investments (by 8.6%); and to lower production costs (by 0.4%).

Conclusions

1. It has been established that in the absence of an acidic catalyst, high-energy particles cause the intensive transformation of vinyl monomers accompanied by the inoculation of radical-active monomer and by its co-polymerization. Among the encountered processes, the formation of inoculated co-polymer predominates. In the presence of an acidic catalyst, together with the general stimulation of resin-forming processes, the content of methylolamides in resin composition substantially increases. In this case, the greatest part of the finishing agent comes into the reaction of cellulose macromolecule linkage, thereby resulting in the formation of cross-links.

2. It has been revealed that the intensification of the processes of resin formation with the participation of a mixture of vinyl monomers and methylolamides is accompanied by a rise in the dye fixation degree in the fabric. The highest yield of fast-fixed dye is noticed in dyeing processes using reactive, direct, vat, and acid dyes (93.6-99.4%).

3. The result of the quantitative evaluation of the solubility of modified cellulose in the cuprammonium solution has proven that under the influence of high-energy particles, the dyes interact not only with thermoreactive resin which is being formed, but also with the macromolecules of cellulose and polyacrylamide. The last of these named reactions greatly changes the distribution of a dye between the resin and the fiber by increasing the portion of dyestuff fixed directly in cellulosic polymer. It must be noted that the radiation-chemistry initiation of the reactions in the system "cellulose - finishing agent - dye" may change the color characteristics of dyed fabric.

4. Highly effective methods have been created for cotton textile material dyeing and finishing using irradiation. It has been shown that the recommended methods (tested in industrial conditions) give the possibility to impart high coloristic indices and consumer properties to shirting which has been dyed and finished according to these methods. Due to the high fixation degree of dyes and finishing agents (89.1-100%), a marked decrease in the quantity of dyestuffs and formaldehyde-containing preparations in waste water is achieved.

References

1. K.D. Pismannic, V.D. Orechov and B.L. Tsetlin, *J. Chem.Soc. named after D.I. Mendeleev,* **XXVI** (4) (1984) 41 (in Russian).

2. J. Gordon, *J. Appl. Polymer Sci.* **5** (18) (1961) 734.

3. E.A. Osminin and B.N. Mel'nikov, in *Collected articles of the Ivanovo Institute of Chemistry and Technology,* **4** (11) (1969) 107 (in Russian).

4. J. Marshy, *Textile Recorder,* **78** (928) (1960) 63.

5. F.I. Sadov, *Laboratory Practical Work in Chemical Technology of Fibrous Materials* (Moscow: Gizlegprom, 1963) (in Russian).

6. In *Analytical Chemistry of Polymers, Vol.1,* Ed. G. Klein (Moscow: Foreign Literature, 1963) 82 (in Russian).

7. J.G. Frick, *Text. Res. J.,* **30** (7) (1960) 495.

8. P. Mehta and R.D. Mehta, *Text. Res. J.,* **31** (2) (1961) 182.

9. In *Reference Book: Finishing of Cotton Fabrics. Pt.1. Technology and Assortment of Cotton Fabrics,* Ed. B.N. Mel'nikov (Moscow: Legprombytizdat, 1991) 409 (in Russian).

10. H.E. Jung et al., *Amer. Dyestuff Rep.,* **66** (3) (1972) 64, 67.

11. T. Batter and D.Y. Treger, in *The Properties of Inoculated and Block-Copolymers* (Leningrad: Khimiy'a, 1970) 54-112 (Russian Edition).

12. G.A. Klein et al., *The Influence of Nuclear Irradiation and Radiation Inoculation in the Fibers* (Moscow: Lyogkaya Promyshlennost, 1968) 110-116 (Russian Edition).

The Institute of Chemistry of Non-Aqueous Solutions of the Russian Academy of Sciences, Ivanovo

INFLUENCE OF MECHANICAL TREATMENT ON THE TECHNOLOGICAL PROPERTIES OF STARCH SIZE

I.M.Lipatova, I.L.Sedova, N.A.Yermolayeva, V.A.Padokhin, A.P.Moryganov

The mechanical treatment of starch paste in a rotor-cavitation activator has been shown to be a new method of size preparation from starch and its mixtures with synthetic polymers. The influence of treatment parameters on size viscosity has been researched. It has also been established that the application of mechanochemically activated size provides sized yarn elongation at break increase for 20-30% which is not accompanied by a decrease in tensile strength.

Starch and its derivatives are the most widely used polymer materials in the Russian textile industry for cotton warp sizing. Being unmodified, starch is inferior to synthetic sizing agents, primarily in film-forming and adherence properties. Nevertheless, starch products cannot be substituted by synthetic materials in sizing processes, because of the insufficiency of raw materials, high cost, and environmental pollution. That is why the problem of starch modification (aimed at the improvement of its technological properties and at augmentation of the purposefulness of starch-containing products) still remains acute. The modification of natural starch can be performed by applying chemical, physical, and fermentative methods as well as by using different combinations of them [1]. Physical methods of starch modification for textile industry application are not yet widely spread, but they are attracting more and more interest. Some examples are: using radio-active irradiation for starch depolymerization [2], the mechanical treatment of starch (by employing a high-speed stirrer) to make its properties stable [3], and the application of ultra-sonic equipment to obtain fine-disperse starch size [4, 5]. The authors of the present article have created and tested some rotor-cavitation pattern activators (RCA) for the mechanical treatment of water-polymer systems. In such activators, the strong shearing forces and ultra-sonic fields (generated by the cavitation) are the main types of physical influence upon the material. In earlier publications, the possibility to use RCA for the regulation of the rheologic properties of polysugar-water systems is shown [6].

The present article is devoted to the investigation of the problems of whether it is possible to apply RCA for starch size production and property improvement and what is the efficiency of such treatment.

Experiment

The properties of the size, obtained by an mechanochemical method, were compared with the properties of the size, obtained by the usual thermo-chemical method according to the following recipe/formula: corn starch - 50 g; chloramine - 0.12 g; NaOH - 0.075 g; water - up to 1000 g. The cooking of the size continued until chloramine was fully destroyed (in about 1.5 hours), in conformity with the procedure usually used at the textile factories.

The mechanochemical method of obtaining size consisted in the following operations: the preparation of corn size dispersion containing 50 g of size and 1000 g of water; heating the system to the paste formation temperature (75-80° C); loading the paste into the activator, and activating for 10-20 seconds.

Cotton yarn sizing has been performed using a laboratory sizing apparatus equipped with a drying unit with an IR-irradiator. The temperature of the impregnation solution was equal to 90° C while drying was carried out at 150-180° C.

The gravimetric method was used to define the amount of paste on the textile material. Physico-mechanical properties of sized yarn were evaluated using the breaking machine PM 3-1 (GOST 66 11 2-73). Before testing, the yarn was conditioned at a relative humidity of 65%.

Results and discussion

The degree of size corn destruction is one of the main effectivity criteria of the starch size preparation method. The starch, which has not been destructed, does not possess film-forming properties and acts as a ballast and thereby decreases the degree of purposefulness of starch products. In the present work, the degree of starch destruction in the size (prepared by the mechanochemical method in which the rotor gyration speed (n) was varied) has been estimated using the centrifugating method [7]. In Table 1, the results are presented of several methods of starch size preparation.

As the table data demonstrate, the mechanochemical method is not worse than the thermal one and exceeds the thermochemical method of size preparation in the degree of starch decomposition.

The partial substitution of starch by a synthetic polymer is one of possible ways of improving starch size film-formation and size adherence properties. But the applicability of such mixtures and also the synthetic component concentration in them are restricted by the incompatibility of starch with the majority of carbo-chain polymers. This incompatibility is a reason for the exfoliation of mixed size (containing more than 20% of synthetic component) in the course of its storage without stirring. The data in Table 1 show that the mechanochemical method of mixed size (starch - polyvinyl alcohol = PVA) preparation ensures a double increase in size stability to the exfoliation. The size appeared to be stable for 7-8 hours, providing the possibility of its utilization within one working day. This result can be explained by the fact that RCA treatment increases the dispersing degree of micro-heterogeneous water-starch systems and, therefore, leads to a substantial enhancement of system homogeneity. Similar results have been obtained on the aggregative stability of starch mixtures with other synthetic polymers.

Table 1

	The method of size preparation		
	Thermo-chemical	Thermal [8]	Mechanochemical
Temperature, °C	98	140 (p = 2.5-3.0 atm)	60-70
Process duration, min	60-80	60	0.2
The decomposition degree (%) of the starch, made of			
corn	75-80	96-98	90.4 (n = 2000 rev/min) 94.5(n = 3000 rev/min)
potato	80-85	98-100	98.5(n = 6000 rev/min) 100 (n = 3000 rev/min)
The stability of the mixture starch : PVA (75:25) to the exfoliation, hours	3-4	–	7-8

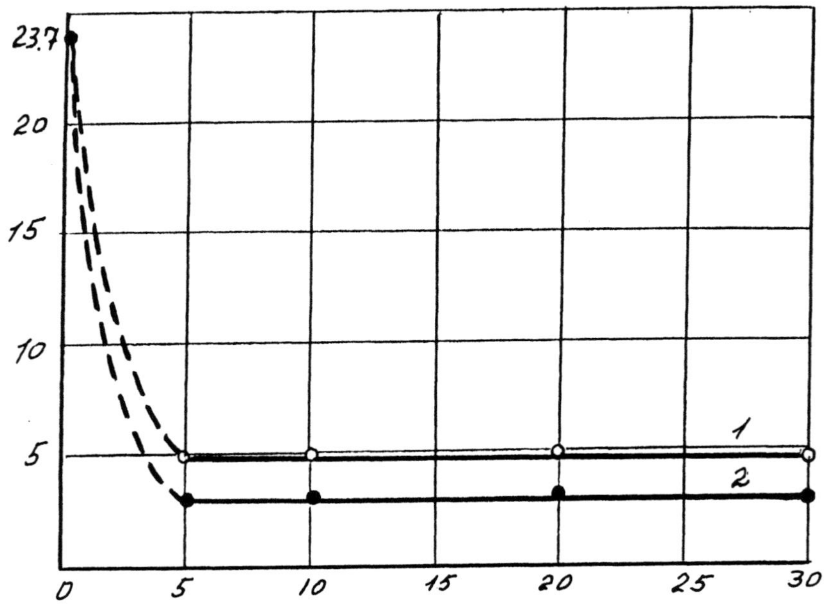

Fig. 1. The dependence of starch solution kinematic viscosity on activation time:
1 - n = 2700 rev/min; 2 - n = 5400 rev/min.
X = τ, s;
Y = $\eta_{rel.}$

The task of paramount importance while creating a new sizing method is the establishment of the correlation between the parameters of the treatment and size viscosity values at sizing temperature (85°C). The treatment of infinite starch paste using RCA causes a large paste viscosity reduction. The dependencies of starch solution kinematic viscosity upon the activation time (for two speeds) are represented in Fig. 1. It is easily seen that during the first 5 seconds, viscosity diminishes to a definite limited value, which does not vary essentially if the treatment duration is further increased. Reasons exist to propose that the maximal paste structure break takes place at the moment of the highest resistance to the shearing, when the system viscosity and, consequently, the expenditures of mechanical energy for the deformation are the highest. Mechanical energy expenditures for structure break are also enhanced when the shearing rate is increased. That is why the limited minimal value of the viscosity diminishes when the speed of rotation is enlarged, as the data in Fig. 2 demonstrate. Thus, the mechanochemical method of size preparation enables the size viscosity to be regulated

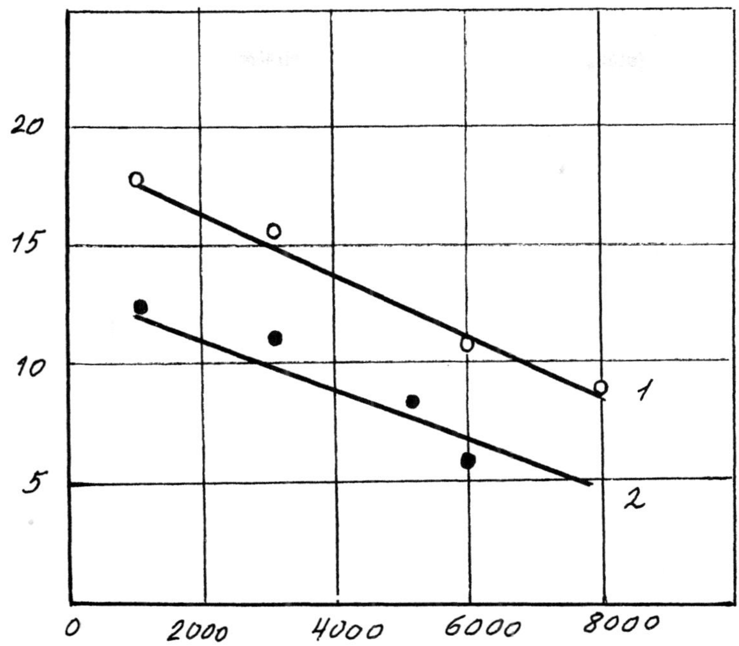

Fig. 2. The dependence of the viscosity of pure starch solution (1) and the solution of starch with the addition of polyvinyl alcohol (2) on rotor gyration speed.
X = n, rev/min;
Y = $\eta_{rel.}$

by the variation of RCA speed conditions. The minimal permitted speed of rotation is defined by the achieved degree of starch decomposition.

Thus, having compared the data, obtained by reologic and microscopic research, one can suppose that the decrease of starch size viscosity under the influence of vigorous shearing strength is, first of all, connected with the destruction of starch paste infinite structure (formed by stuck together and swelled grains) and with the further demolition of starch grain covers.

The effectivity of the mechanochemical method of size preparation has been evaluated according to the quality indices of the size and sized yarn. The data obtained are represented in Table 2 together with the data characterizing the effectivity of the thermochemical method of yarn sizing.

Table 2

Yarn	Size composition	Thermochemical method				Mechanochemical method			
		$\eta_{rel.}$	Paste, %	$\sigma \cdot 9.8 \cdot 10^{-3}$ N	Δ l, mm	$\eta_{rel.}$	paste, %	$\sigma \cdot 9.8 \cdot 10^{-3}$ N	Δ l, mm
N 34	Corn starch	16.8	10.0	342 ± 62	12.5 ± 2.4	7.9	8.2	341 ± 60	17.5±2.3
	Starch : PVA (90:10)	16.5	10.6	331 ± 60	13.9 ±2.4	7.1	8.3	340 ± 61	16.1±2.5
	Soft yarn	–	–	248 ± 49	20.7 ± 2.2	–	–	248 ± 49	20.7±2.2
N 50	Corn starch	16.8	12.5	220 ± 38	10.6 ± 1.8	8.9	13.8	233 ± 42	11.3±1.6
	Soft yarn	–	–	166 ± 35	17.1 ± 2.3	–	–	166 ± 35	17.1±2.3

The analysis of the results lead us to the conclusion that the positive influence of size mechanochemical decomposing agents is displayed mostly in sized yarn elongation at break increase (Δl). If compared with inactivated size, this increase is equal to 40% when pure starch is applied and to 20% when a starch - polyvinyl alcohol mixture is used for sizing (particularly yarn N 34). The enhancement in tensile strength (relative to the soft yarn one) in both cases is the same and equals (on average) 35% (σ). These data correlate with those obtained earlier, which have revealed that water-starch solution mechanical treatment (using RCA) enlarges the elasticity of the films made from these solutions [9].

As appears from the above, the possibility in principle exists of starch paste mechanical treatment which can serve as an independent method of fine-dispersed size preparation. This method is an alternative to the traditional thermochemical and thermal ones. The advantages of the new method are the following:

- the exclusion of chemical decomposing agents from the sizing composition and, as a result, the lowering of environmental pollution;
- the elimination of the cooking process which requires large-scale equipment and high energy expenditures;
- the provision of a high homogeneity of many-component size;
- the improvement of the technological properties of size.

References

1. M. Vill, *Der Alimentation*, **93** (1981) 70, 72.

2. S. Bachman, S. Witkowski and M. Pietka, *J. Rad. and Nud. Chem. Lett.*, **118** (3) (1987) 185.

3. Patent Application 58-76475, Japan..

4. Patent Certificate DE 3530770, Germany.

5. P.G.Dumitrash and Yu.N. Paukov, *Thesis of the report on the 10-th Rumanian Conf. on textiles and leather* (Yassy, Rumania, 14-16 of May, 1992) 19.

6. I.V Nuzhdina et al., *Thesis of the report on the 1-st All-Union Conf. "Liquid-phase materials"* (Ivanovo, 15-20 of Oct., 1990) 133 (in Russian).

7. *Reference book on cotton weaving,* Ed. P.T. Bukay'ev (Moscow: Legprombytizdat, 1987) 153 (in Russian).

8. I.V Bykova, N.V. Volkova, in *Collected articles: New in technique and technology of cotton fabric finishing* (Moscow, 1982) 13 (in Russian).

9. I.M. Lipatova, I.L. Sedova and A.P. Moryganov, *Thesis of the report on the Internat. Conf. "Textile chemistry"* (Ivanovo, 26-29 of Oct., 1992) ? (in Russian).

The Institute of Chemistry of Non-Aqueous Solutions
of the Russian Academy of Sciences, Ivanovo

A NOVEL TECHNOLOGY FOR PRODUCING TEXTILE MATERIALS WITH A PROLONGED MEDICAL EFFECT

G.E.Kritchevskiy, L.B.Savilova, N.D.Oltarzhevskay'a,
A.Ya.Polishchuk, A.L.Iordanskiy

A novel method for producing dressings and bandages with predefined medical effect is proposed which consists of the introduction of a medicinal preparation (MP) into a composition prepared on the basis of bio-compatible polymers and of the subsequent putting of this composition onto the textile material using printing techniques. The general model of MP transfer within the system "textile bearer - polymer film - thickener" is established and analyzed. The understanding of MP desorption and polymer thickener dissolution mechanisms, thus obtained, makes it possible to forecast and regulate the kinetics of MP release owing to variations in the boundary conditions (polymer film solubility, MP solubility, etc.).

The development of novel textile goods for medical purposes with a prolonged medical effect is a fascinating and significant trend, especially concerning dressings for the mass treatment of wounds and burns in every day medical and home practice and in extreme situations. The traditional production of such materials is a complicated chemical and technological process, expensive, and requiring special demands on the chemical nature of the fiber and MP [1, 2].

However, a novel, new method of producing dressings with a prolonged medical effect has been developed [3]. The method consists of the introduction of MP into the composition, prepared on the base of bio-compatible polymer gels, and in putting this composition onto the textile material using textile printing techniques. These techniques permit the bringing onto the surface of a material any medicine, even hardly dissolved, which is impossible to be brought in high concentrations by any other method (for example, by impregnation). The choice of MP is defined by its compatibility with the polymer and by medical prescription [1, 2].

The application of printing techniques for the achievement of this new goal of producing goods for medical purposes requires the polymers in the composition to be bio-compatible and to possess their own biological activity and medical influence. That is to say: the role of these polymers must not be restricted by the function of the MP "bearer" but they must serve as "pro-medicine", promote the prolongation of MP desorption into the outer medium (a wound), and have reological and printing characteristics suitable for the application of printing techniques. Such polymers as polysugars (alginates, carboxymethylcelluloses, dextran, polyvinyl alcohol, chitosan, etc. and their mixtures) belong equally to the province of medical men and textile chemists alike [4].

To create materials with predefined medical properties and with a prolonged effect, it is important to define MP diffusion and desorption rates from the polymeric textile material and from the polymer film layer (covering the material's surface) into the wound. In this case, the wound must be presented as the outer medium of small volume (liquor ratio = 3÷5). This process is similar to unfixed disperse dye desorption during the rinsing of printed fabrics, providing that stirring in the external medium is absent. The concentration of medicine in the wound will increase due to its transfer out of the thickener layer and simultaneously decrease due to the preparation's inactivation and its transfer into the blood and lymph in the wound. As a result, the outflow of the preparation will take place. Such a situation may be defined as the changing of diffusing substance concentration in a limited volume, accompanied by the outflow of the substance through the polymer layer surface if the substance flow is not stationary [5]. To do this, it is necessary to describe the regularities of medicine transfer into the wound, to clarify the criteria for the choice of the composition, to evaluate the optimal structure of the textile material bearer in order to use all possibilities and advantages that are given by the application of the textile printing method in order to produce materials with predefined medical properties.

But the question is how to achieve the prolongation of MP outflow into the external medium (wound) and how to regulate the duration of this process while applying printing techniques.

It is experimentally proven that polymer gel located on a textile surface slows down the transfer of MP into the outer medium and creates a prolongation effect. Thus, the use of a printing method gives the possibility to obtain medical textile materials with a prolonged MP outflow into the wound.

As is evident from Fig. 1, the manner of MP printing and, consequently, the manner of MP distribution in the textile material influences the persistence of the medical effect.

Later, we will prove that the MP distribution in the dressings is as follows:

–one part of MP is located on the surface of the polymer film. It is soluble in water (and hence in the exudation) and serves as a "shock dose" necessary for treatment;

–another part of MP (70-75%) is spread inside the polymer film volume. The outflow rate of this part of MP depends upon its properties (for example, solubility), the physical and chemical characteristics of film forming polymers, and the properties of the external medium (The ooze liquid coming from the wound was simulated by a mixture of polyelectrolytes, each in a concentration of 0.1 g/l. This mixture had a pH of 7.4.);

–the third part (20-25%) is sorbed by the textile bearer and is retained by adherence forces.

The usual demands made of these dressings are the presence of a "shock dose" of medicine in the first moment of treatment and a constant rate of MP transfer into the wound for 2-3 days thereafter [2]. The heightened MP concentration in the surface layer of the film is defined by the conditions of manufacture including drying during which MP is diffusing to the film surface together with water. The existence of an MP "shock dose" equal to 5-7% of the total MP content is proven in [5, 6] with the help of special procedures based on simulating the actual environmental conditions in which MP takes effect. These procedures allowed us to quickly make an evaluation of MP desorption. Dressings printed with a composition containing MP were immersed into a liquid (simulating the ooze from a wound) of volume V_0 (liquor ratio M < 10). After the

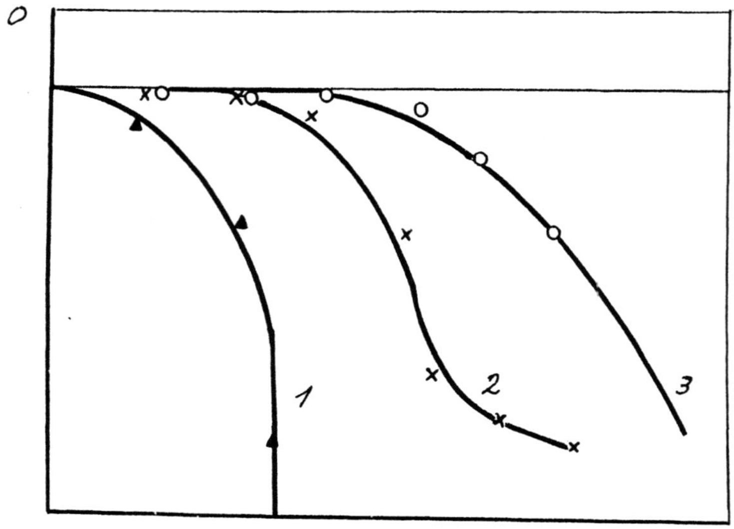

Fig.1. The prolongation of the medical effect of nitrofurazone (liquor ratio < 10, the external medium is partly substituted) by means of impregnating fabric in an alcohol solution (1), and of nitrofurazone printing in the composition with either CMC (2) or manutex (3).
X = *t* (time, without units);
Y = *c* (concentration in the solution, without units).

definite period of time $\Delta\tau$ had passed (2, 5, and 10 minutes), a definite part of the total volume (00.5; 0.1; 0.2 V_O) was substituted by an equal volume of initial liquid without medicine. The portions of liquid taken away were analyzed for their MP content. The relative MP outflow rate was defined according to the formula:

$$w = \frac{\Delta V}{V_0 \Delta t}.$$

In reality, the MP inactivation rate in human organisms equals $w = 5 \cdot 10^{-5}\ s^{-1}$. The advantages of the described procedures consist in the possibilities to analyze the MP desorption rate much more quickly and to govern the experimental desorption rate by a variation of *w*. In Fig. 2, it is shown that at the initial stage ($t < 1/w$) MP concentration in the external medium is a linear function of time (MP - nitrofurazone, $w = 3.63 \cdot 10^{-4}\ s^{-1}$), but the straight lines do not pass the origin of the coordinates. This regularity is connected with the presence on the outer surface of the polymer film of some quantity of MP which is flowing out rapidly into the external medium (wound). It is exactly this MP

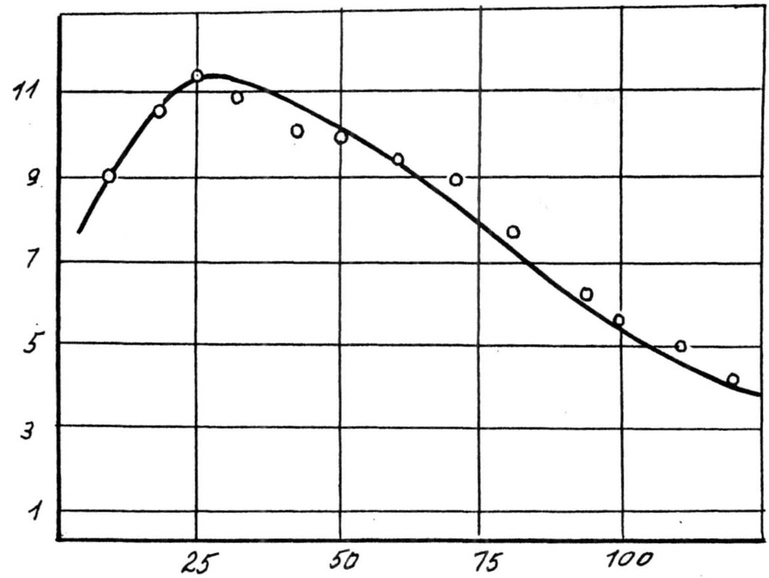

Fig.2. The kinetic curve of MP desorption (liquor ratio < 10, the external medium is partly substituted).

X = t, min

Y = c, μg/ml.

quantity which provides the shock medical effect and may be qualified as the first part of the preparation.

The existence of the second and third MP parts is proved by the kinetic curves of alginate and nitrofurazone desorption into the external volume (Fig. 3). These curves were obtained using a spectrophotometric method. To make alginate noticeable in the solution, it was first dyed in a concentrated solution of sulfuric acid. [7]. After the experiment had been completed, the content of dyed alginate in distilled water (liquor ratio M = 200) was estimated. It goes without saying that the real conditions of dressing application are much more complicated, but in the experiment this method of approximation was adopted.

The kinetic curves in Fig. 3 demonstrate the difference between the dissolution rate of alginate polymer put on the surface of the textile bearer and the rate of MP outflow. The continuation of nitrofurazone desorption after the complete alginate dissolution confirms our supposition about the presence of MP not only in the polymer film (second part) but also in the textile bearer (third part). In order to clarify the mechanism of transfer of

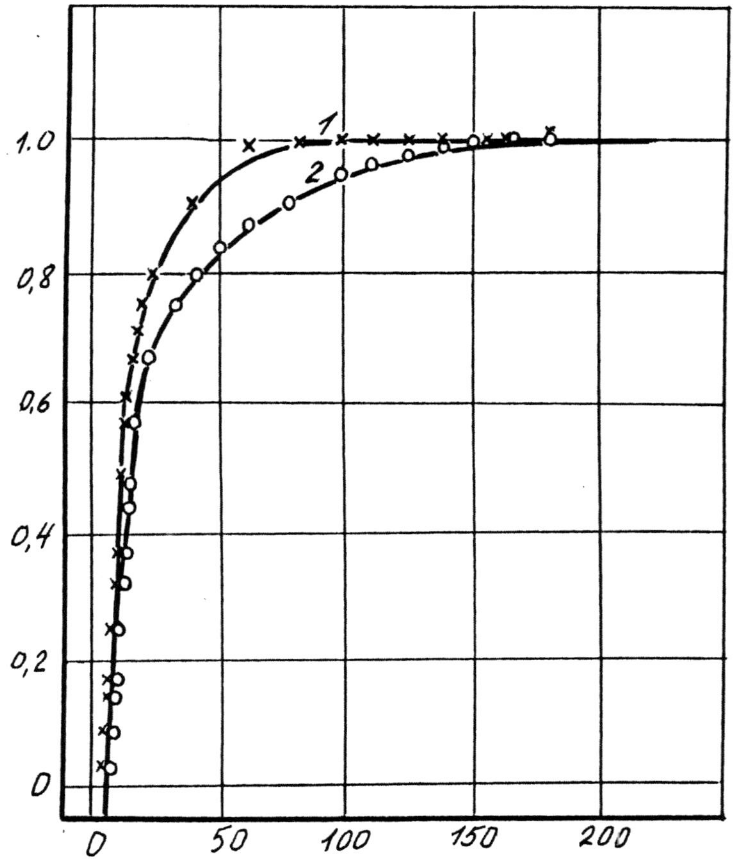

Fig.3. The kinetic curves of alginate (1) and nitrofurazone (2) desorption.
X = t, min
Y = $c_d/c_{d\infty}$.

MP's second and third parts, a comparative analysis was made (for different ways of putting MP on the surface of dressings) of the kinetics of MP outflow into the external medium. In the first series of experiments, MP was included in the printing composition with polymer. In the second, MP was brought onto the textile substrate by means of padding with the MP aqueous dispersion of the same concentration as in the printing paste. The MP desorption from the padded pattern simulated the process of the outflow of the third MP part.

Research into the kinetics of polymer-thickener accumulation in the external solution (Fig. 3, curve 1) reveal that the following relation is the best analytical approximation:

$$c_p = c_{p\infty}\left(1 - e^{kt}\right),$$

where c_p is the current thickener concentration in the solution; $c_{p\infty}$ is its limiting concentration in the solution, corresponding to the complete polymer film dissolution; k is the dissolution rate constant; and t is the time of film dissolution.

As the thickener accumulation in the external solution reflects the film dissolution process, the experimental results indicate the possibility to describe the thickener dissolution by the first order equation:

$$\frac{dV_p}{dt} = -kV_p; V_p(0) = V_{p_0},$$

where V_p is the current and V_{p0} the initial polymer-thickener volumes on the surface of the textile bearer.

Having been limited by the supposition that the polymer film cross-area is constant, one can receive the following expression for the film thickness (l) alterations:

$$l = l_0 e^{-kt}, \quad (1)$$

where l_0 is the initial film thickness.

If the noted discreteness of the thickener distribution on the textile material surface is borne in mind, then l acquires a physical meaning of the mean polymer film thickness.

Because the second MP part is located inside the polymer film, the premise can be made that MP outflow is connected with polymer swelling and with the dissolution rate. The kinetics of the dissolution of bio-compatible polymers-thickeners, brought onto the textile material, in the external medium (i.e. distilled water) was estimated by using high pressure liquid chromatography at a small liquor ratio. Among the polymers, the following were chosen: sodium alginate with a molecular weight of 30,000-50,000 and 150,000-200,000; a mixture of sodium and calcium alginates (manutex RS in which the presence of the calcium salts provides: first, the formation of salt cross-links between the carboxyl groups of the polymer macromolecules and secondly, an increase in thickening power); and the water-soluble sodium salt of carboxymethylcellulose (CMC). In conformity with the rate of the outflow into the outer volume (M=5), the encountered polymers are arranged in the following order: alginate > manutex >CMC. It has been established that alginate with a molecular weight of 30,000-50,000 possesses the greatest outflow rate. In all the experiments conducted, the quantities of the investigated polymers brought onto a material surface of equal area were taken as 100%. The results are represented in Table 1.

Table 1.

Polymer	$C_p / C_{p\infty} \cdot 100\%$ at the time (t ,min)						
	10	20	30	40	50	60	70
Manutex RS	30.0	81.8	86.4	100.0	100.0	100.0	100.0
CMC	13.8	34.7	58.3	76.3	100.0	100.0	100.0

According to the rates of swelling and of dissolution of the polymer films which were founded on purpose in distilled water at a liquor ratio of 40 (these rates were evaluated indirectly using the data concerning an increase in the solution viscosity), the polymers under discussion form the following sequence: alginate > manutex > CMC. The same regularities are maintained when the polymers swell in blood plasma [8].

The comparison of the polymer swelling rate and the kinetics of the preparation desorption into the external medium reveals the existence of the correlation between these indices. The alginate outflow rate ($1.90\ 10^{-3}\ s^{-1}$) is greater than the rate of CMC outflow ($1.20\ 10^{-3}\ s^{-1}$). The lower the molecular weight of alginate, the higher is the desorption rate of MP: the use of alginate with a molecular weight equal to 30,000-50,000 gives the desorption rate of $1.90 \cdot 10^{-3}\ s^{-1}$ while the desorption rate of alginate with a molecular weight of 130,000-150,000 reduces the speed of desorption down to $0.63 \cdot 10^{-3}\ s^{-1}$. An increase in the polymer-thickener concentration in the composition leads to a decrease in the MP mass-transfer rate.

The physical structure of a polymer substantially influences the processes under investigation. For example, the introduction of a shrinking agent (glutaric aldehyde) into the composition deaccelerates MP desorption from the polymer film. The higher the aldehyde concentration and the longer the duration of treatment, the lower is the MP outflow rate.

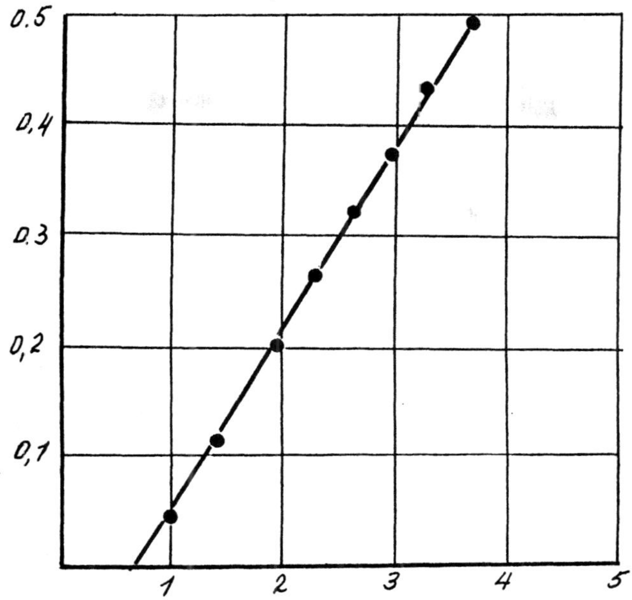

Fig.4. The kinetic curve of MP desorption.
X = $\sqrt{t}$, min$^{1/2}$;
Y = $c_d/c_{d\infty}$.

The results obtained prove that the rate of outflow of the second part of MP depends on the swelling and desorption rates of the polymer in which the preparation is distributed.

At the same time, the difference between the polymer outflow rate and the rate of freeing the second MP part points out the influence of the preparation transfer proper on its desorption rate into the external volume. The supposition about the diffusional mechanism of MP transport inside the polymer seems to be the most reasonable explanation and is supported by the experimental data (Fig.4). As is seen from Fig.4, at small periods of time corresponding to the constant initial thickness of a film ($t << 1/k$), the concentration of the preparation in the solution rises in direct proportion to $\sqrt{t}$ and thereby confirms the correctness of our former reasoning. The experimental results allow the calculation of the MP diffusion coefficient in the film as:

$$\frac{\partial c_d}{\partial t} = D_d \frac{\partial^2 c_d}{\partial x^2} \quad (2)$$

$$0 < x < l_0$$

with the boundary conditions

$$c_d(l_0, t) = 0, \ \frac{\partial c_d}{\partial x}(0,t) = 0, \ c_d(x,0) = c_{d_0}, \quad (3)$$

where c_d, c_{do} = the current and initial MP concentrations in the polymer film; D_d = MP diffusion coefficient. It should be noted that the condition used for the boundary "polymer - textile bearer" ($\frac{\partial c_d}{\partial x}(0,t) = 0$) is valid only at the first stage of outflow when the variations of MP concentration on this boundary, due to the transfer of the third part of the preparation, can be disregarded.

The solution of the equation (2) with respect to MP concentration in the solution (c_d) gives the expression:

$$\frac{c_d}{c_{d_\infty}} = 2\sqrt{\frac{D_d t}{l_0^2}},$$

where c_{d_∞} = the limiting MP concentration in the solution. It is the input parameter for the calculation of the reduced diffusion coefficient (D_d/l^2) whose values for different preparations are presented in Table 2. The application of the reduced diffusion coefficients is conditioned by experimental difficulties while defining l_0.

The data thus obtained form the basis for the quantitative simulation of the desorption process of the second MP part as a diffusional transfer in a polymer film of variable thickness. However, in order to describe the kinetics of the outflow of two MP parts (second and third) as a whole (this procedure is expedient because the separation of these two parts is conventional to a marked degree), it is necessary to consider the alterations of MP concentration on the boundary "polymer - textile bearer".

The results demonstrate that in the absence of a polymer film, the outflow of the third MP part is best described by the following equation:

$$c_d = c_{d_\infty}\left(1 - e^{-kdt}\right).$$

This equation gives the possibility to presuppose that the process of freeing MP from the textile material is limited by the first order chemical reaction:

$$\frac{dc_d}{dt} = -k_d\left(c_{d_\infty} - c_d\right)$$

With respect to MP concentration in a textile material (c_d), this equation is as follows:

$$\frac{dc_d}{dt} = -k_d c_d ; c_{d(0)} = Kc_{d_0} ,$$

where K = MP distribution coefficient between a bearer and a polymer-thickener. The remainder of the processes connected with MP transfer inside the bearer may be disregarded. The values of outflow rate constants of the third part of the preparations of nitrofuran series (characterized by a similar molecular structure but by different molecular weights and solubilities), together with the values of D/l^2 for the second MP part (Table 2), show the significant influence of MP solubility upon the rates of outflow of both second and third MP parts. The reduction of MP solubility results in the diminution of the rate of its outflow into the external solution.

Table 2.

MP	M	Chemical structure	Solubility in water, %	Desorption rate	
				D_d/l^2	K
Nitrofurazone	198	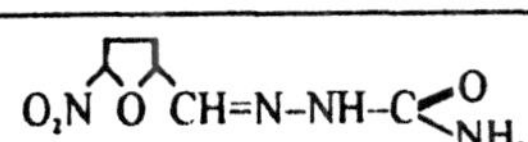	0.02000	4.30	4.10
Furadonine	238	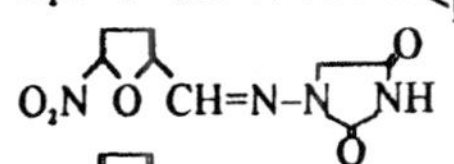	0.01250	1.96	1.62
Furagine	264	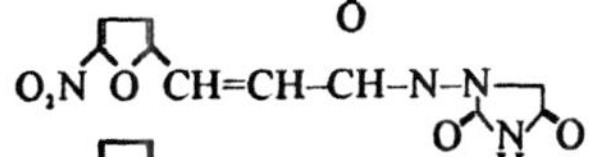	0.00770	1.13	0.32
Furazolidone	225	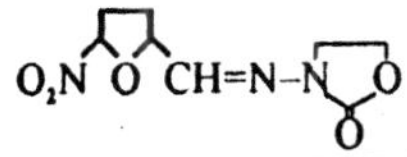	0.00400	0.86	0.46

The generalized model of MP transfer in the system "textile bearer - polymer-thickener film" describes three simultaneous processes: the dissolution of the polymer

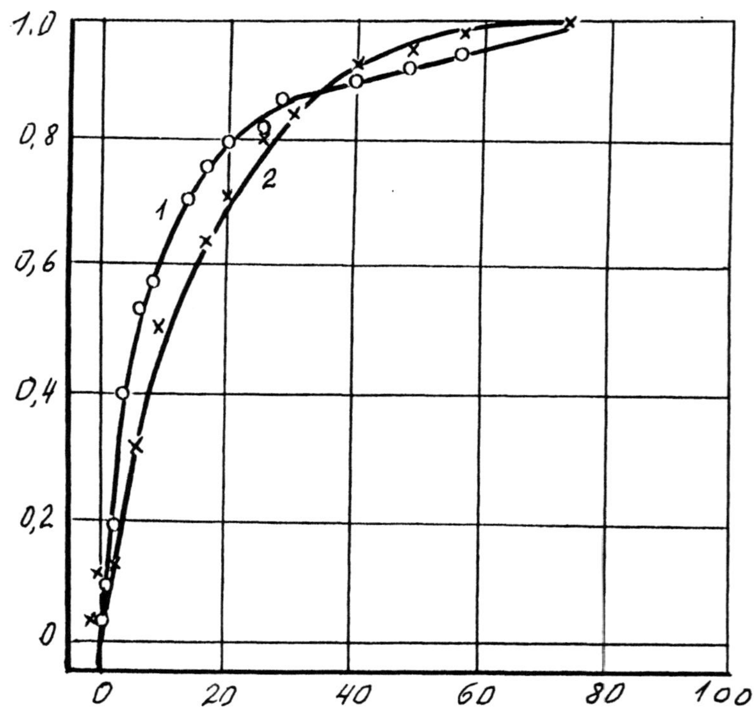

Fig.5. The calculated (1) and the experimental (2) curves of MP desorption.
X = t, min
Y = $c_d/c_{d\infty}$.

film (eq.1), the diffusion of MP inside the polymer film (eq.2), and the freeing of MP out of the textile material in the absence of the polymer film (eq.4). By solving the generalized system of these three equations in the following boundary conditions:

$$c_d = kc_d(0,t); c_d(x,0) = c_{d_0}; c_{d(0)} = Kc_{d_0},$$

one can define the parameters k and K (in the absence of a polymer film). This opens the possibility to estimate quantitatively MP desorption into the external medium at any moment of time. The excellent coincidence between the calculated and the experimental data (Fig.5) is indicative of the adequate description of the process of MP diffusion out of dressings by the proposed generalized model.

The ideas obtained of the mechanism of MP desorption and of the polymer-thickener dissolution together with the revealed physicochemical regularities of the discussed processes open the possibility to predict and to regulate the kinetics of MP outflow by means of the alteration of the boundary conditions (polymer film solubility, MP solubility, etc.).

Thus, the application of the industrial textile printing method (used for obtaining patterned fabrics for a long time) for the new purpose of creating dressings with a prolonged medical effect results in a 1.2-1.5 time reduction in the expenditure of medicines, the avoidance of traumatisation during bandaging due to the presence of polymer-thickener gel in the wound, and the simplification of the technology for the production of dressings.

References

1. A.E. Vasily'ev, *Itogi Nauki i Tekhniki. Ser.: Khimiya i Khimicheskaya Tekhnologiya Visokomolekulyarnykh Soyedineniy,* **16** (1988) 3 (in Russian).

2. M.M. Fel'dshtein et al., *Khimiya i Khimicheskaya Tekhnologiya Visokomolekulyarnykh Soyedineniy*, **16** (1988) 120 (in Russian).

3. Patent Certificate 1643612, USSR.

4. N.D. Oltarzhevskay'a et al., in *Collected Articles of the Scientific Research Institute of Textiles and Notions,* Ed. V.N. Filatov (Moscow: Publishing Agency of the Central Scientific Research Institute of Light Industry, 1988) 36 (in Russian).

5. T.S. Vasily'eva et al., *Izv.Vuzov. Tekhnol. Tekstil. Prom.,* (1) (1991) ??? (in Russian).

6. T.S. Vasily'eva, O.A. Subbotko and I.A. Grigiry'eva, in *Collected Articles of the Scientific Research Institute of Textiles and Notions,* Ed. V.N. Filatov (Moscow: Publishing Agency of the Central Scientific Research Institute of Light Industry, 1989) 30 (in Russian).

7. N.I. Kochetkov, *The Methods Used in the Chemistry of Carbons* (Moscow: Mir, 1967) (in Russian).

8. L.B. Proshina et al., in *Collected Articles: The Researches into the Creation of Textile Materials for Medical Purposes,* Ed. V.N. Filatov (Moscow: Publishing Agency of the Central Scientific Research Institute of Light Industry, 1990) 73 (in Russian).

The International Engineering Academy,
Moscow

THE INFLUENCE OF AQUEOUS AMMONIA SALT SOLUTIONS ON THE CONTENT OF THE OLIGOMERS ON THE SURFACE OF DYED POLYESTER FIBER

N.P.Prorokova, S.M.Prusova, S.Yu.Vavilova, Yu.A.Kalinnikov

The influence of aqueous ammonia and aqueous ammonia salt solutions upon the content of cyclic oligomers on the surface of dyed polyester fiber has been investigated at different ammonia and salt concentrations, temperatures, and treatment durations. The role of ammonium chloride in the compositions studied has been identified. The optimal concentration, temperature, and time parameters of the process for treating polyester fiber with aqueous ammonia solutions have been defined. These parameters ensure the maximal decrease in the oligomer content on the fiber surface.

In the Institute of Chemistry of Non-Aqueous Solutions of the Russian Academy of Sciences, a range of novel liquid-phase materials has been created which enable the substantial intensification of discontinuous polyester dyeing processes [1]. These multi-component materials include water, ammonia, and salts. Therefore, the necessity has arisen to study the dependencies of polyester fiber parameters upon the indices of fiber treatment in aqueous ammonia and aqueous ammonia salt solutions. One of the most interesting questions is the problem of the changes in cyclic oligomer content on the polyester fiber surface, especially because those oligomers that are located in the surface layer hamper the further processing of polyester and worsen the quality of polyester goods [2].

In the present research, such polyester fiber was used which had been dyed with disperse dyes in industrial conditions but not subjected to a reductive treatment. The fiber was treated with aqueous ammonia or aqueous ammonia salt solutions in hermetic, stainless steel dyeing cones at 30-135^{0}C for 1-35 minutes. The ammonia concentration in the solutions varied from 0 up to 1.5 g/l. The oligomer content on the polyester fiber

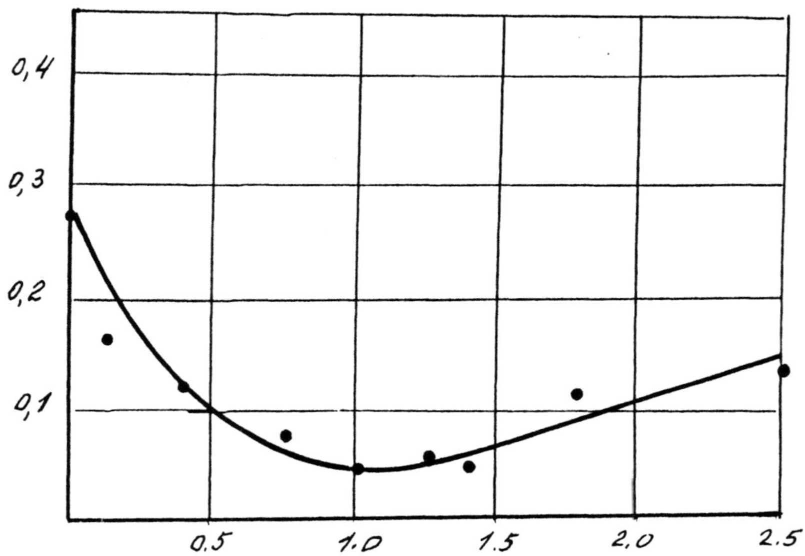

Fig.1. The dependence of the oligomer content on the surface of polyester fiber upon the concentration of ammonia in aqueous ammonia solutions.

X = C_{NH_3}, g/l;

Y = Oligomer concentration on the polyester fiber surface, %

surface was evaluated spectrophotometrically (using a Specord M 40 spectrometer) in accordance with the well-known procedure described in [3].

The dependence of the oligomer content on the surface of polyester fiber, treated with aqueous ammonia, on the ammonia concentration is presented in Fig.1. Fiber treatment was performed at 130^0C for 5 minutes. From the data in Fig. 1, it can be seen that an increase in the ammonia concentration of up to 1 g/l causes a significant (6.5 times) decrease in the amount of oligomers on the fiber surface (region A). Successive ammonia additions into the solution lead to a certain rise in the oligomer content on the fiber (region B). If the data obtained in [3] are born in mind, the described dependence can be interpreted as a result of two simultaneous chemical reactions. The first reaction is a low-mass (<1000) cyclic oligomer (presumably trimer) hydrolysis provoking the dissolution of these oligomers and, consequently, the diminution of their content on the fiber. This reaction predominates in region A. The second reaction is the destruction of high-mass (>1000) oligomers which are difficult to separate from the polymer. This reaction predominates in region B.

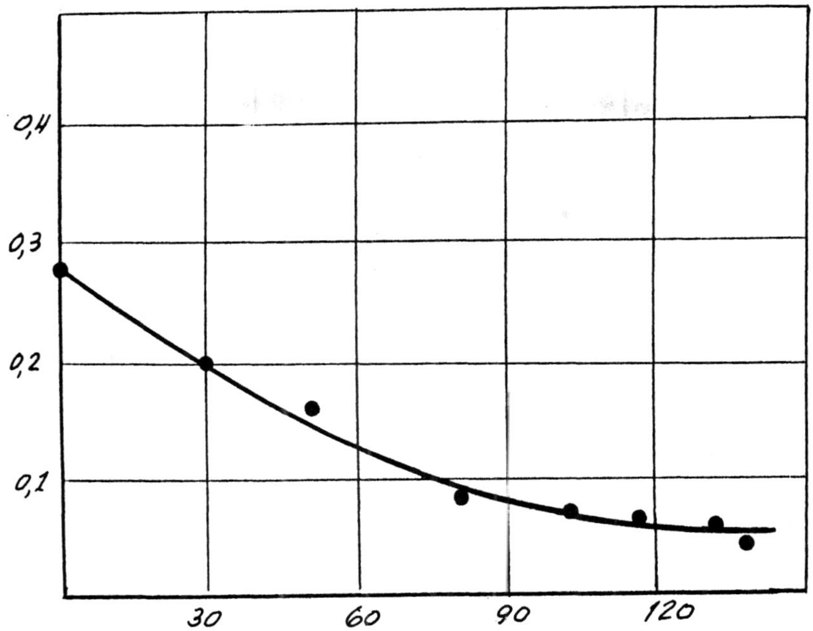

Fig.2. The dependence of the oligomer content on the surface of the polyester fiber upon the temperature of fiber treatment in an aqueous ammonia solution.
X = temperature,^{0}C
Y = Oligomer concentration on the polyester fiber surface, %

The data in Fig.2 illustrate the dependence of the surface oligomer content upon the temperature of a 5 minute polyester treatment with a solution containing 1 g/l of ammonia. As follows from Fig.2, the temperature rise promotes the reduction of the oligomer content on the fiber surface. However after the temperature reaches 100^{0}C, a further temperature increase does not cause an essential decrease in the oligomer content. This phenomenon may originate in the intensification of the oligomer migration from the inner fiber zones towards the fiber surface.

The dependence of the surface oligomer content upon the duration of polyester treatment with an aqueous solution of ammonia (1 g/l) at 130^{0}C is demonstrated in Fig.3. The data show that maximal reduction of the oligomer content is observed after 5 minutes of fiber treatment. A subsequent increase in treatment duration does not produce a diminution of the oligomer content. Obviously, in the conditions cited, an equilibrium is achieved between the processes of dissolution of the oligomers being hydrolyzed and their migration from the depth of the fiber towards its surface.

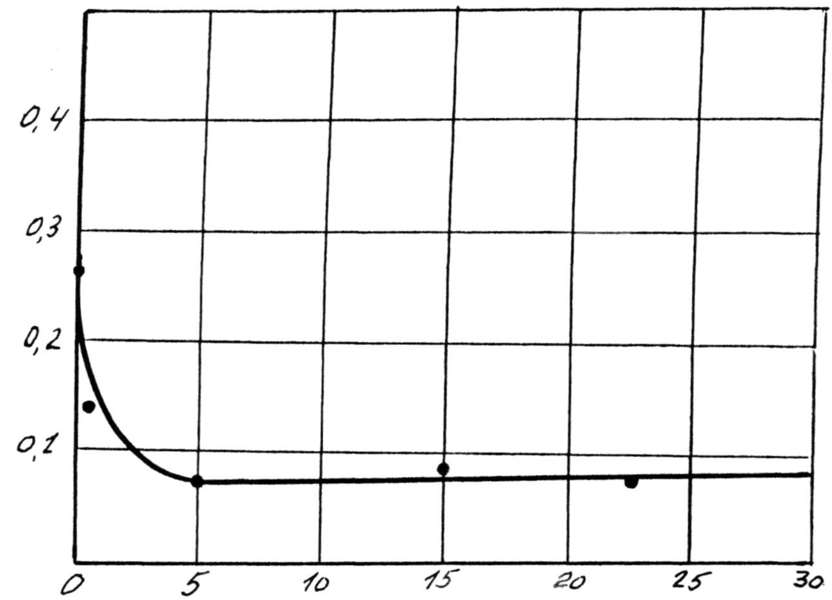

Fig.3. The dependence of the oligomer content on the surface of the polyester fiber upon the duration of fiber treatment in an aqueous ammonia solution.
X = time (τ), min;
Y = Oligomer concentration on the polyester fiber surface, %

The introduction of one of the salts (used in the new intensive dyeing technology previously proposed by the authors in [1, 2]) into an aqueous ammonia solution changes the character of the dependencies that have been just discussed (Fig.4). The graphs presented in Fig.4 reflect the influence of ammonia (curve 1) and salt (curve 2) concentrations in aqueous ammonia salt solutions upon the oligomer content on the fiber. The polyester treatment was performed at 130^{0}C for 30 minutes with two types of solutions: one with a constant ammonium chloride concentration (0.5 g/l) and a variable ammonia concentration (curve 1); and the other with a constant ammonia concentration (0.5 g/l) and a variable ammonium chloride concentration (curve 2).

The comparison of the dependencies in Fig.1 and Fig.4 (curve 1) leads to the conclusion that the addition of ammonium chloride into the aqueous ammonia solution slows down the rate of hydrolysis of both low-mass and high-mass oligomer fractions. The absence of a B-region (which is easily noticed in Fig.1) on the curve in Fig.4 is indicative of a large decrease in the rate of reaction 2. The deceleration of reaction 1 is confirmed by an increase in the remnant content of cyclic oligomers on the fiber surface.

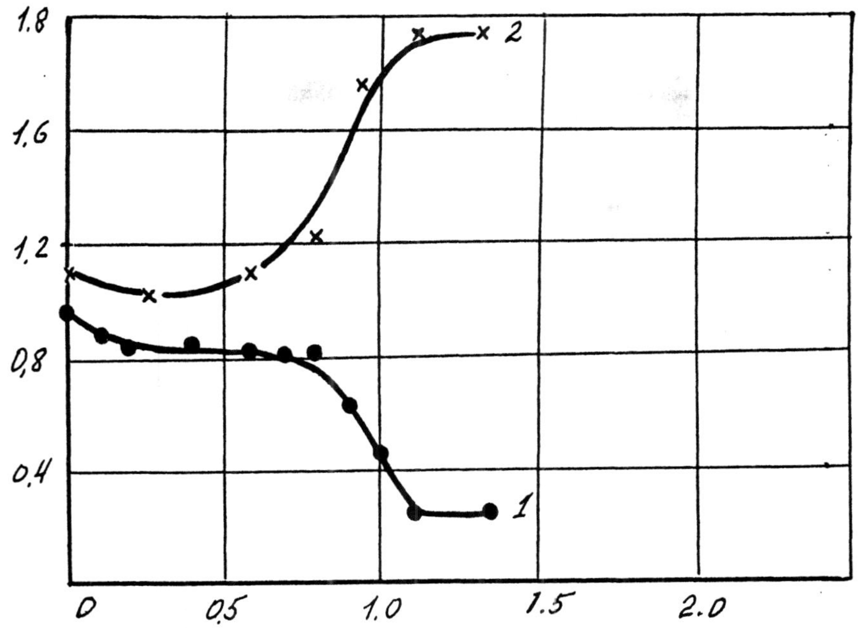

Fig.4. The dependence of the oligomer content on the surface of polyester fiber upon the concentrations of ammonia (1) and ammonium chloride (2) in aqueous ammonia salt solutions.
X = *c*, g/l;
Y = Oligomer concentration on the polyester fiber surface, %.

This content equals to 0.06% after fiber treatment in the aqueous ammonia solution and 0.21% after fiber treatment with the aqueous ammonia salt solution (in which the concentration of ammonium chloride was 0.5 g/l). When the ammonium salt concentration in the solution is raised (Fig.4, curve 2), an augmentation in the oligomer content takes place, being defined (due to the suppression of the reaction 1) only by the amount of the oligomers that migrate from the inner fiber layers towards the surface.

Thus, it is reasonable to conclude that in the course of polyester fiber treatment with aqueous ammonia, ammonium salts perform a defense function impeding the fiber from hydrolysis under the influence of ammonia. At the same time, ammonium salts do not permit the effective refinement of the fiber surface by the removal of cyclic oligomers. As for the aqueous ammonia solutions, they enable the almost complete removal of the oligomers from the fiber. That is why it is expedient to perform a separate operation of polyester cleaning by rinsing with an aqueous ammonia solution. This operation, based on the newly established effect, can be easily included into any known discontinuous industrial process of polyester dyeing.

Generalizing the data in Fig.1-4, it is possible to define the optimal parameters of polyester fiber treatment with aqueous ammonia solutions. These parameters ensure the maximal reduction of the oligomer content on the fiber surface and are: an ammonia concentration of 0.75-1.25 g/l; a 5 minute duration of the treatment process; and a temperature range of 100-110^{0}C. The data obtained give the possibility to create a new highly effective method for the simultaneous cleaning and rinsing of polyester fiber.

References

1. Patent Certificate 1617072, USSR.
2. D.L. Shan, *Man-made Text. India*, **24** (2) (1981)8
3. *The Analytical Control in the Production of Synthetic Fibers (Reference book)*, Ed. A.S. Chegoli, and N.M. Kvasha, (Moscow: Khimiya, 1982)148 (in Russian).

The Institute of Chemistry of Non-Aqueous Solutions
of the Russian Academy of Sciences, Ivanovo

THE INFLUENCE OF OXIDIZING AGENTS ON LINEN FABRICS DURING THE FIGURE BLEACHING PROCESS

E.V.Vladimirtseva, L.V.Sharnina, I.B.Blinitcheva, B.N.Mel'nikov

The nature of the influence of different oxidizing agents (hydrogen peroxide and sodium bensylsulfochloramide) on lignin and flax cellulose during the process of local bleaching of coarse linen-containing textile material has been determined. Linen fiber destruction has been estimated at those places where the design has been applied, provided that high oxidizing agent concentrations in the printing paste have been used. It has also been established that both the achieved degree of the design whiteness and the stability of the fabric fastness properties ensure the high effectivity of the created figure bleaching process.

Textile materials made of flax are very popular all over the world. Usually bleached linen fabrics are dyed and printed with the gray linen fabric assortment remaining untouched.

A new, original method for figure bleaching of gray linen textile goods has been developed at the Ivanovo Academy of Chemistry and Technology in the Department of Fibrous Material Chemistry and Technology. This method differs from the particularly well-known ones in the procedure of getting the design. The design appears due to the local destruction of coarse linen colored admixtures, not due to dye printing. The bleaching agent is the main component of the printing paste. This agent concentration is several times greater than in the traditional bleaching bath (in liquid-phase bleaching process) and is varied according to the degree of design whiteness required. The duration of the influence of the bleaching composition on the textile material equals several minutes [1].

The present work was aimed at examining the changes taking place in lignin and flax cellulose under the influence of different oxidizing agents during the local bleaching of flax-containing textile materials. Those substances, usually applied in linen-fabric pretreating processes, were chosen as oxidizing bleaching agents: hydrogen peroxide and sodium bensylsulfochloramide (chloramine B).

Gray linen fabrics served as the objects of this research. The fabric materials chosen differed in their structure and domestic purpose (Table 1).

The design whiteness measurements were performed at the spots of the application of the compositions with different bleaching agent concentrations. A Spekol-11 spectrophotometer has been used with the measurements being taken at the light absorption wave length $\lambda = 540$ nm. The results obtained are represented in Fig. 1.

Table 1.
Quality characteristics of flax-containing materials

Article of the fabric	Compositi on	Assortment	Thickness , mm	Surface density, g/m^2	Impurity	Botton lightness of the initial fabric, %
411	flax	for clothing	0.29	150	unclogged	30.7
07114	flax	for laundry goods	0.38	213	slightly clogged	31.7
10252	flax:chem . fiber (67:33)	for technical buckram	0.58	280	consider-ably clogged	24.5

The changes in flax lignin in the course of bleaching were evaluated with the aid of UV-spectroscopy. As an example, several spectra of linen fiber dioxane extracts are shown (Fig. 2), the fabric having been bleached with different bleaching agents. The first spectrum group (Fig. 2 - a) characterizes unbleached linen fiber (taken from the fabric of the article 07114) while the second (Fig. 2 - b) characterizes the same fiber after solid-phase bleaching. Identical correlations have also been obtained with the other textile materials.

The band, located at 275-290 nm, is distinctly seen in the spectra of infinite pattern lignin extracts. Such a picture is typical for native lignin. This band is known to originate from the existence of an aromatic structure closely connected with chromophoric group

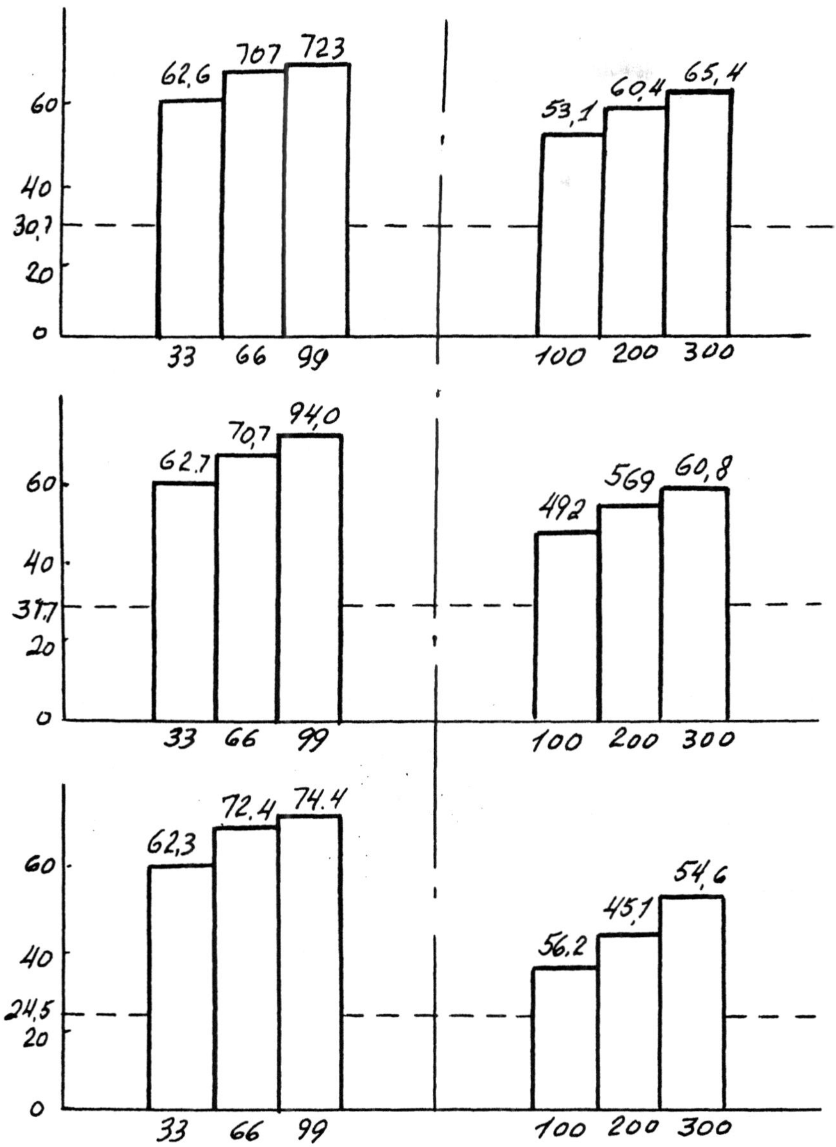

Fig. 1. Whiteness degree of the fabrics treated according to the solid-phase bleaching process:
(a) - article 411; (b) - article 07114; (c) - article 10252;
I - the composition, containing hydrogen peroxide in the concentration (g/kg):
white field - 33; hatched field - 66; double-hatched field - 99;
II - the composition, containing chloramine B in the concentration (g/kg):
white field - 100; hatched field - 200; double-hatched field - 300; dashed line - initial fabric botton lightness.

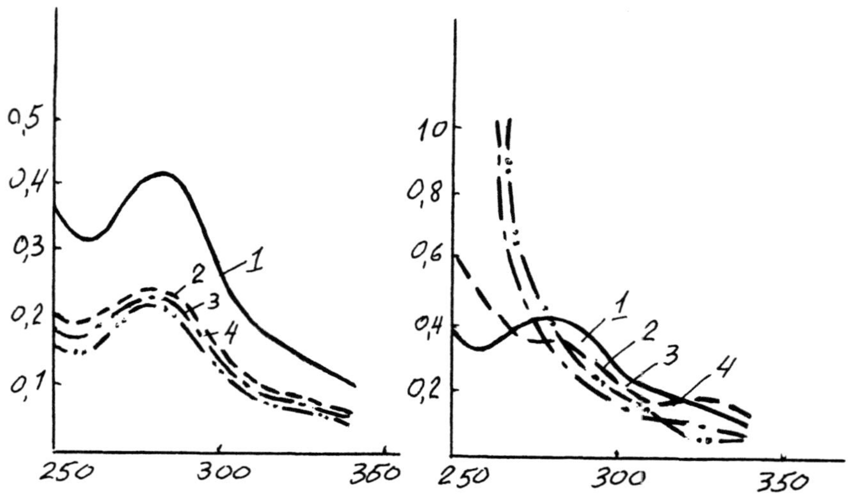

Fig. 2. UV- spectra of lignin dioxane solutions obtained by lignin extraction from the fabric (article 07114):
(a) - hydrogen peroxide-containing composition:
1 - initial fabric; 2, 3, 4 - bleached fabric (the concentration of bleaching agent is equal to 33, 66, 99 g/kg, respectively);
(b) - chloramine B-containing composition:
1 - initial fabric; 2, 3, 4 - bleached fabric (the concentration of bleaching agent is equal to 100, 200, 300 g/kg, respectively).
X = λ, nm; Y = Light absorption of lignin solutions (optical density).

formation, responsible for gray flax color [2]. It is usually considered that the main reason of lignin natural coloring is the presence of chromogenic functional groups (bonded to benzene rings by conjugated chemical bonds) in the molecules of lignin. The most important of these groups are: carbonyl groups (conjugated with the main molecule body), double bonds in benzene ring side chains, and the combination of these two chromogens [3].

During the influence of printing paste (containing hydrogen peroxide) on the linen material, a total decrease of light absorption intensity all over the spectral interval studied is observed. This is indicative of the structural link destruction between lignin molecules and of lignin partial removal out of the fiber. However, the absorption band at 275-290 nm still remains in residual lignin spectra despite the high degree of linen material whiteness achieved. This means that the hydrogen peroxide bleaching action (in the procedure described) is conditioned upon the destruction of the chromophoric groups (ketones, double bonds, etc.) while the aromatic lignin bodies remain almost untouched.

If chloramine B is used as the bleaching agent, the spectral curve shape changes greatly. The absorption band, which corresponds to aromatic lignin macromolecule structure, almost disappears, giving an evidence of the destruction of this structure. Obviously, the cause of the destruction consists in lignin oxidization and chlorination processes, resulting in chinoid group formation. The yellowish shade of the design, obtained after the treatment of the fabric with chloramine paste, confirms this supposition.

While bleaching, the influence of bleaching agents affects not only lignin, but also the cellulosic component of the stalk fiber. As was mentioned earlier, in figure bleaching processes, the bleaching agent concentration on the fiber possesses "shock" value in comparison with the concentration in the traditional bleaching process. That is why our research has been carried out to evaluate the degree of flax cellulose damage. For this purpose, such indices as cupric-ammonium solution viscosity and textile material tensile properties have been defined (Table 2).

Table 2.
Tensile properties of linen textile materials bleached (according to the figure bleaching process) with compositions containing hydrogen peroxide (1), or chloramine B (2)

Tensile properties	Fabric article					
	411		07114		10252	
	1	2	1	2	1	2
Cellulose damage degree, %	0.34*	0.40	0.01	0.10	0.90	0.95
Losses in tensile strength (%), corresponding to the area of fabric hiding of						
50%	-7.4**	1.2	-2.9	-7.3	-0.9	3.2
100%	3.4	4.4	-9.4	-10.4	2.2	1.9

* - the admissible cellulose damage degree is equal to 0.4% [4]

** - the unprinted fabric, having been subjected to the same wet and thermal treatments as the printed one, served as a pattern for the comparison

The table data demonstrate that solid-phase bleaching does not worsen textile material tensile properties. The degree of cellulose damage exceeds admissible values only when a low-grade yarn fabric (article 10252) is treated.

The comparison of the degree of cellulose damage caused by the application of different bleaching agents reveals that hydrogen peroxide is a less strong destructing agent than chloramine B. This conclusion correlates with the data obtained from UV-spectroscopy.

Thus, our research data leads us to conclude that figure bleaching technology, in which "shock" bleaching agent concentrations are used, constitutes a new effective coloring process for gray linen textile materials.

References

1. Patent application 93-034429, Russian Federation.
2. F.E. Brauns and D.A. Brauns, *The Chemistry of Lignin* (Moscow: Lesnaya Promyshlennost, 1963) (Russian Edition).
3. I.A. Lebedeva and S.P. Kretchetova, *Wood-pulp Bleaching* (Moscow: Lesnay'a Promyshlennost, 1973) (in Russian).
4. *Reference Book on the Chemical Technology of LinenP Fabric Treatment* (Moscow: Lyogkaya Promyshlennost, 1973) (in Russian).

The State Academy of Chemistry and Technology,
Ivanovo

INDEX

P

R

S

T

U

V

W

Z